D1741019

God's Other Son

SIR ISAAC NEWTON

1642-1727

God's Other Son

ISAAC NEWTON

The Link Between Science and Religion

James A. Brettell

Writer's Showcase
presented by *Writer's Digest*
San Jose New York Lincoln Shanghai

God's Other Son
The Link Between Science and Religion

All Rights Reserved © 2000 by James A. Brettell

No part of this book may be reproduced or transmitted in any form or by any
means, graphic, electronic, or mechanical, including photocopying, recording,
taping, or by any information storage retrieval system, without the
permission in writing from the publisher.

Writer's Showcase
presented by *Writer's Digest*
an imprint of iUniverse.com, Inc.

For information address:
iUniverse.com, Inc.
620 North 48th Street, Suite 201
Lincoln, NE 68504-3467
www.iuniverse.com

ISBN: 0-595-12957-9

Printed in the United States of America

Acknowledgements

When I was about eight years old, a family friend, Juel Burkhardt, observed my prowess with simple arithmetic and was impressed that I could solve multiplication problems in my head. He told me that I should read about Isaac Newton—the great mathematician. I remembered his advice all my life. But, like so many other people, making a living for my family became my first priority and consumed all my time. It was not until I retired that I recalled his advice and realized that now I had the time to read about Isaac Newton.

I began the project merely as an intellectual exercise for my own pleasure and did not count on it becoming an all-consuming effort. The more I read, the more fascinated I became with Newton, the people he referred to as the "giants," and those whole lives he touched. Seemingly unrelated facts suddenly presented themselves to me which aroused my interest even more, and researching those facts often revealed a remarkable relationship, or a coincidence, when considered in relation to other facts and the times in which they occurred.

The great number of apparent coincidences in the histories of religion and science which appear to be related and that plead for recognition and explanation is the basic purpose of this book. And the relationship between the events described is my perception of the story of the universe.

I have made a conscious effort to limit the length and complexity of this book in order to simplify mathematical information and explanations for the common reader. To the many authors from whose research my knowledge has been largely derived, I express my thanks and appreciation for their scholarly achievements; most of the material contained in this book is the work of those authors. My perception of the relationship of God to the other people mentioned and His purpose for putting Isaac Newton and all the "giants" on this earth is my only claim to originality and is the basic purpose for sharing this point of view.

I shall be forever grateful to my wife, Joan, for encouraging me to make the trip to Cambridge to complete my research and to experience, first-hand, the magnificence of the libraries of Trinity College—Newton's alma mater, King's College, the Isaac Newton Institute, and of course, Cambridge University. Her assistance in transcribing notes from the unpublished, handwritten, Newton's manuscripts was extremely helpful. She graciously tolerated my many hours of isolation with my nose in a book while I researched this material and developed my theory of the relationship between science and religion.

All my original contacts with Cambridge University were via e-mail so that by the time we arrived all necessary arrangements were established. Everyone in Cambridge was extremely cordial and went the extra mile to assist me in my quest for information. Adam Perkins greatly assisted us in expediting our admission and establishing our reader status for access to the magnificent resources of the University libraries. Scholars everywhere will forever be indebted to Peter Jones at King's College. He single-handedly edited the Newton manuscripts—given to the College by John Maynard Keynes—in order to reduce them to microfilm; this expanded the manuscript's availability to students of the Newton mystique. Without these manuscripts on microfilm, Newton's unpublished religious writings would have been lost to scholars forever. Allison Sprostin at the Trinity College library—where Newton spent so much of his time—was very knowledgeable and helpful and provided confirmation of a sort of reverence about the classical Christopher Wren structure that still stands so perfectly proportioned and elegant after more than 300 years. Original letters written by Isaac Newton and various artifacts on display serve to emphasize the solemnity of the entire structure. Heather Hughes and

Andrea Le Core of the Isaac Newton Institute greatly assisted me in establishing priorities to make my time in Cambridge more productive.

I am very grateful to the authors of books listed in the bibliography, especially Richard A. Westfall, Frank E. Manuel, and Matt Goldish, for their interest, diligence, and unique ability to express their understanding of the life of the greatest scientist the world has ever known. Their works establish the standard to which all other literary efforts about Isaac Newton, his science and his religion, must be measured.

The ultimate confirmation of Isaac Newton's contributions to science was provided by Gregory Oliver, Chief of the Ascent and Descent Dynamics Office, Barbara Conte, Group Leader of the Ascent Analysis Branch, and Phillip Burley of the Flight Dynamics Office at the Johnson Space Center in Houston, and Dr. Steven Dutczak, Chief of the Education Department at the John F. Kennedy Space Center. Every launch of every space vehicle utilizes Newton's Laws of Gravity and Laws of Motion. And they all still work perfectly.

This endeavor would not be nearly so understandable had it not been for the outstanding editing efforts of Lucia Dulin Hawkins. She discovered errant words and phrases that would have escaped without her diligent efforts, and her organizational suggestions were also very appropriate and very much appreciated.

WREN LIBRARY

TRINITY COLLEGE

CAMBRIDGE UNIVERSITY

Contents

Contents

Preface

God's displeasure with the inhabitants of this war-plagued planet in the years preceding the birth of Jesus was first evidenced by His decision to create The Great Flood and select Noah to re-populate the Earth. However, wars and killing continued; cruelty was common, and slavery was the fate of the defeated. It was no wonder that God became disillusioned with the conduct of His people in this land where crime was rampant. Although certainly justified in taking drastic action, perhaps God made a mistake when He destroyed the Tower of Babel and confused the language of the people to prevent the building of the tower.

Not until Sir Isaac Newton emerged in the mid-seventeenth century have all the people of the Earth again spoken the same language. Perhaps God tired of waiting and hoping for the people to work things out for themselves when He finally sent Jesus to teach virtues, noble principles, moral codes, and ethics for their everyday life. With His son teaching and the disciples spreading the word, God surely believed that some of the new people would discover how the universe worked. No one did. After more than 1600 years, He possibly tired of waiting, and just as He put Jesus on this Earth, He also put Isaac Newton here on December 25, 1642. But the common birthday is not the only similarity between the two. Although there is no evidence of a virgin birth, Newton was born three months after his father's death in the midst of the great Civil War in England. Even though we know very little about his early years we do know that, like Jesus, his most notable achievements occurred by the time he was 24 years old. Isaac Newton discovered the spectrum of light, the laws of motion, and the laws of gravity.

And Isaac Newton invented calculus. Through this universal language that everybody could understand, he explained to the people of the world how practically everything in the universe operates. He calculated the force necessary for an object to break the bonds of gravity and go into orbit around the Earth, and he calculated the familiar buttonhook path space shuttles follow to return to Earth. It was Isaac Newton who explained the gravity-speed-centrifugal force relationships necessary for the orbit to take place. Newton's greatest work, "The Principia" is indisputable evidence of his incredible perception and understanding of an extremely wide range of subject matter. Written in less than two years, it was essentially complete before he was 24 years old. The entire Principia is written in calculus terms—the universal language every person in the world can understand, regardless of his native tongue.

Although Edmund Halley, of Halley's Comet fame, was 14 years younger than Newton, I like to think of them as contemporaries and close friends. I imagine Ed calling Ike, his familiar name for Isaac, and saying something like, "Ike, I have this theory about the comet and if you'll show me how to do this calculus thing you've invented, I believe I can compute the path and predict the return of the comet," and Ike saying, "Sure Ed, come on over." We all know that Halley did, in fact, predict the return of the comet, but it is also important to know that the comet returned on the birthday of Jesus Christ and Isaac Newton in 1758, long after both Newton and Halley were dead. Perhaps it was a birthday present from God as gratitude for fulfilling His wishes to tell the world how the universe operated. It appears to be more than a mere coincidence.

Newton was convinced, beyond the shadow of a doubt, of the existence of God and said that things were too perfect to have happened by chance. He believed that the planets had to be distributed in a certain manner by an initial act before the principal of gravity could come alive. Newton also said that comets were phenomena in whose progress God had to intervene from time to time. He believed that there had been incidents of major cosmic or geological catastrophes in the past and that re-population of the earth after such an occurrence required a divine degree. From his adolescence onward, the pursuit of knowledge as power over things and of knowledge as a revelation of God was ever present in his mind.[1]

Even a cursory study of Isaac Newton's life reveals his intense efforts to put everything in order where no loose facts or circumstances are unaccounted for or left unexplained. He wanted everything to be absolute. Not satisfied with the unexplained beginnings of ancient kingdoms, he tried to establish the chronology of crucial historic events in order to determine dates absolute for such events as the Argonautic Expedition and the fall of Troy.[2]

Newton believed that every discovery of a scientific principal and every correct reading of a prophetic text demonstrated the goodness and orderliness in the universe God created. When he wrote the "Principia", he said that he hoped it would help men to believe in a Deity. He said that it was the religious duty of the scientist, who was capable of unraveling the wonders of God's creation, to reveal them to mankind, and if the scientist failed in this task, he denied God one form of adoration.[3]

Newton eagerly recognized the findings of scientists' discoveries before him said that, "If I have seen further than other men, it is because I have stood on the shoulders of giants".[4]

It has been said over and over by many different people that Isaac Newton was the greatest man who ever lived. Voltaire paraphrased one of his letters in a proposed epitaph for Newton agreed and said that, "It is for him who masters our minds by the force of truth, and not those who enslave them by violence, that we owe our reverence."[5]

---

1 Frank E. Manuel, <u>A Portrait of Isaac Newton</u>, p.124

2 Frank E. Manuel, <u>A Portrait of Isaac Newton</u>, p.165

3 Ibid, p.126.

4 Richard Westfall, <u>The Life of Isaac Newton</u>, p.106

5 Voltaire, <u>Letters Concerning the English Nation</u>, Letter XII

Chapter 1

Original Religion

The Bible leaves very little to doubt regarding the origins of religion. Because years were counted differently when it was written, any expression of time is uncertain; however, according to the Bible, about 1600 years after God created Adam the world was so filled with violence and corruption that He decided it could no longer continue like that. Wars were commonplace. With everyone attacking and with the victor subjecting the vanquished to the most repulsive treatment and conditions imaginable, no doubt God watched in horror at these abominations until he became completely fed up with the situation and announced that He intended to destroy all the people. After careful consideration He selected Noah—the tenth generation born after Adam and the grandson of Methuselah—and told him he was the only one worthy to re-populate the earth. God told Noah He intended to make it rain for forty days and forty nights to flood the earth and proceeded to tell Noah exactly how to build an ark. At this time Noah had no children and he was 480 years old. We are told that Noah lived three hundred fifty years after the flood and that his fourth generation descendant, Nimrod, became a successful warrior whose lands extended to large areas of Mesopotamia. Although his identity is debated, the Book of Genesis describes Nimrod as "the mighty hunter," which is interpreted to mean that his prey was man. He was a lawless and impious tyrant who built his fortress on the basin of the Tigris and the Euphrates Rivers in what we know as Babylonia. And so it appears that the cycle of wickedness, violence, terror, and corruption of man began all over again.

It is difficult to accurately set the year, but we know that one hundred years after the great flood only one religion and one language existed. The Book of Genesis in the Old Testament records this fact and tells of the people who came to the basin of the Tigris and the Euphrates Rivers to settle and build a city and construct a tower to the heavens.

The city we know as Babylon and the tower as Babel. God did not like this idea because He believed that the people had forgotten Him and were beginning to worship their own creations. This appeared to be an open, daring sin and rebellion against God. He said that their pride was sufficient sin and did not want the people to be capable of doing just anything they wanted. He said, "Look, they are one people and they all have one language; and this is only the beginning of what they will do; nothing that they propose to do will be impossible for them." He then confused their language so they could no longer understand each other's speech, and He then scattered them over the face of the earth.

While this idea did not hold great prospects, in retrospect it appears that God may have made a mistake by taking away their language and scattering these people all over the face of the earth. In any event, these circumstances produced even more corruption and violence than before the great flood. Finally, the King of Babylon went too far by attacking Israel and imprisoning and persecuting many Jews. This greatly angered God, who then vowed to destroy everything and everyone in Babylon. During the next few hundred years the people were scattered all over the world and developed and learned new languages related primarily to their geographic region, and in this period a great deal of progress occurred in the development of the intellectual well-being of the people. Although the Babylonians incurred the wrath of God in the years before the birth of Christ, they contributed greatly to science and were very active in mathematics; we owe our system of time and its system of units based on sixty to them.[6] They developed a system of cuneiform characters, stick and wedge-like figures, to record astronomical events and we now know that they were able to predict the movements of the moon with a high degree of accuracy. In addition, they also developed tables of standard mathematical functions such as multiplication, division, squares and square roots, cubes, reciprocals, and many other mathematical procedures.

Other major civilizations, such as Chinese, Maya, and Egyptians were also well informed about the position and movement of the planets, the sun, the moon, and the stars and achieved notable and sometimes amazing scientific discoveries. While these people worshipped a great number of gods, it appears that the common thread to religious beliefs everywhere remained a special sanctity for the four cardinal directions.

The Egyptians built the pyramids around 2500 B.C. and achieved extremely accurate positioning aligning the cardinal directions. Shafts venting to the outside surfaces of the pyramids align with the positions of Thuban and Alnilan—the two stars which Egyptians associated with vegetation and rebirth. Constructed without the aid of any machinery or iron tools of any kind, the building of the pyramids still remains one of the wonders of the world—and a great mystery.[7]

There are many other prehistoric stone monuments, such as Stonehenge in Britain, the standing stones of Calinish on Lewis Isle off the west coast of Scotland, and Newgrange in Ireland. All share a common interest with the cardinal directions, the observation of the sun, the moon, and the stars and suggest the possibility of astronomical alignments and perhaps the predictions of eclipses. Such predictions, of course, were invaluable to the medicine men who spent great portions of their life keeping the monarchy convinced of their importance.

Many great men, most of whom lived in the Middle East and Egypt, worked and studied all their lives to further the cause of knowledge in the

6 Evan Hadingham,Early Man and the Cosmos, p.10
7 Ibid, p.22

world. Pythagoras, who lived about 500 B.C., is recognized by everyone who has taken basic geometry, and the theorem that bears his name is still the standard for determining the lengths of the sides of a right triangle.

Plato, the philosopher, followed Pythagoras and lived from 428 to 347 B.C. In contrast to Pythagoras, Plato's interests also included politics, and he is best known for his academic achievements. Aristotle, one of Plato's students, became a teacher in Plato's Academy in Athens. And at almost this same time period, 365 to 300 B.C., Euclid grew up in what is now Alexandria, Egypt. He is best known for his magnificent treatise on geometry, "The Elements." Also during this same period, 356 to 323 B.C., Alexander the Great of Macedonia, a student of Aristotle, became one of the first educated rulers. Trained in literature, science, medicine, and philosophy, he ruled vast territories and built many cities. Archimedes followed, chronologically, and his contributions centered mainly on mathematics. His methods appear to form the basis for the integral calculus that Isaac Newton invented 2,000 years later.

Eratosthenes, 276 to 197 B.C., became the first librarian at the great library that Alexander the Great built in Alexandria. Eratosthenes estimated the diameter of the earth with surprising accuracy, and he also measured the tilt of the earth's axis. Hipparchus, who is best remembered for his division of a circle into 360 degrees, followed in the period 180-125 B.C. These individuals are evidence of God's confidence that education might be the key to world peace and tolerance. Perhaps God was encouraged enough with the apparent intellectual progress taking place that He reconsidered His position relative to permitting people to know everything they could know. Obviously, the fiasco in Babylon showed Him that keeping people ignorant prevented the development of world peace. During this period, God placed Buddha and Confucius in India and China. This period is what Karen Armstrong, in her book "History of God" called the Axial Age[8] when God delegated responsibility to the world leaders of religion, in languages they understood, to be the guardians and teachers of their people's moral well being. These civilizations—each with a common language—developed independently within each geographical region. With no commercial contact between these eastern regions and Europe, the concept of God in the Far East remained solely based on the philosophical beliefs and teachings of Confucius and Buddha.

All of these people form an uninterrupted succession of intellectual contributors to the world, and the earliest of these are referred to by Isaac Newton as "giants": the brilliant scientists who preceded him. A more elaborate discussion of these giants is presented in the following chapter. Even though great strides in academic achievement occurred and scientific advances made by these great mathematicians and philosophers continued, fighting, corruption, violence, and religious pressures persisted for hundreds of years. Wars raged across Europe, Asia, India, and south into North Africa. Barbaric massacres, persecution of Jews, and the physical ravages of fighting resulted in chaos so severe that God possibly wearied of the constant feuding. Realizing the great flood and Noah's re-population of the earth had not worked out the way He intended, I think God may have thrown up His hands in despair and exclaimed, "I've had it with you people!" Continual fighting and wars among His people defied His most fervent wish for His people to live together harmoniously. Something else had to be done. His people needed someone stronger and with more influence than the Philosophers and the mathematicians; His people needed to be taught ethics, love, justice, and a moral code to create an atmosphere of peace, trust, integrity, and honesty. God believed the momentous decision to put Jesus on this earth to teach the people how to live together, to be the answer to the problem. The name "Jesus" is derived

8 Karen Armstrong, <u>A History of God</u>, p.27

from the Greek interpretation of the Hebrew name Joshua, and "Christ" is derived from the Greek Christos; the anointed one. Early followers of Jesus used the name "Christ" for him because they regarded him as "the redeemer of humanity." While there is some information written by Saint Matthew and Saint Luke about Jesus' childhood, the Gospels are silent concerning his life between the ages of 12 to 30. The Gospels then record the public ministry of Jesus beginning after the imprisonment of John the Baptist and lasting for approximately one year. The Gospel, according to John, indicates that it began with the choosing of his disciples and lasted for about three years.

Jesus promised pardon and eternal life for sinners, no matter the seriousness of their crimes, and he preached the infinite love of God for the weak and the humble. Intellectuals from the time of Pythagoras, 500 years earlier, wrote all forms of documentation of their theories and beliefs, and it is unusual that Jesus depended on other people to record his teachings. Because several of his disciples wrote differing accounts of the same event, we are left tour own devices to decide which is the true recording. Nevertheless, the life and teachings of Jesus have been and continue to be the subject of debate among religious scholars, but his influence on the Christian world is acknowledged as one of the great achievements of all time.

Chapter 2

The Giants

"If I have seen further than other men, it is because I have stood on the shoulders of giants."[9] These are the words Isaac Newton used to refer to the many astronomers, mathematicians, philosophers, and academicians who preceded him who had contributed and recorded significant understanding of virtue, wisdom, reasoning, mathematics, and astronomy. Several of these giants were referenced briefly in Chapter 1 and more detail on their achievements will be provided in this chapter. Most of the giants, however, came after the birth of Christ.

After witnessing the resulting chaos that continued after the destruction of the Tower of Babel, God evidently concluded that ignorance is the parent of anarchy—a state of society without government or law—and therefore changed His opinion about permitting the people to possess knowledge. Newton agreed that ignorance was a not good policy because he said that he believed it to be a religious duty of the scientist, who was capable of unraveling the wonders of God's creation, to reveal them to mankind. To fail in this duty would be to deny God one form of adoration.[10]

---

9 Richard Westfall, <u>The Life of Isaac Newton</u>, P. 106
10 Ibid, p 126

The Tower of Babel incident is considered one of the most significant events in history when it is viewed in relation to the events that followed. Perhaps the single most important and influential result of the incident was the inability of the people to communicate and understand one another. In light of history up to the present day, we have still not fully recovered. Certainly the lack of understanding, differing perspectives and points of view of a given situation, event, or moral interpretation, continue to impede the progress toward peace in the world. Perhaps God made a mistake by taking their language and scattering the people all over the face of the earth. Wasn't that the beginning of the different religions, languages, customs, and rituals that still divide the people of this planet?

After the people were scattered all over the world for the next several hundred years, they developed and learned new languages which related primarily to the geographic region in which they lived. During this time a great deal of progress transpired in the intellectual development and well being of the people. Although Newton doesn't tell us who the giants were specifically, certainly the individuals listed here are among them.

PYTHAGORAS[11]

Born 500 years before Christ, Pythagoras, whose methods may have been known to the Babylonians 1000 years earlier, will always be best remembered for the Pythagorean Theorem which still applies today for solutions to right triangles. He believed that everything could be reduced to number relations and was perplexed to discover that the diagonal of a square is not a rational multiple of its side, but the result confirmed the

---

11 "Pythagoras" <u>World Book Encyclopedia</u> Vol. 15 p 813

existence of irrational numbers. Also being an astronomer, he believed and taught that the earth is a sphere at the center of the universe. He also recognized that Venus as an evening star is the same planet as Venus as a morning star. While he did not quite get to the real working of the system with regard to the relative positions of the earth and the sun in our universe, his discoveries were the first step to solving the riddle of the workings of the solar system.

BUDDHA AND CONFUCIUS

The passive and non-violent doctrine of Buddhism is the identifying characteristic in the political history of Buddhist countries. Buddhism is open to all, regardless of race, and the teachings are basically related to individual conduct. Their "Four Noble Truths" are that human life is unhappy, the cause of this unhappiness is human selfishness, the individual selfishness can be corrected, and the method of correction is the "Eightfold Path"; right views, thought, speech, action, livelihood, effort, mindfulness, and meditation.[12] Confucianism, too, concentrates on individual moral standards of thought and conduct. Benevolent concern for one's fellow man and personal conduct that exercises a combination of manners, ritual, custom, etiquette, and propriety are the virtues on which Confucianism is based.[13] God's delegation of responsibility to these two individuals in their respective regions of the world, and the virtues professed by them, is evidence of God's concern for the welfare of the world and His intent to instill a desire for people to live in peace and harmony.

---

12 Michael H. Hart, The 100, p 23
13 Ibid, p 28

These individuals greatly affected the lives of many people of the world, and as a theology student they vitally interested Newton.

PLATO[14]

Plato, who lived to be 80 years old, is considered to be one of the fathers of Western political philosophy and ethical thought. Born in Athens of a distinguished family, in his early years he intended to pursue a career as a politician. However, as a young man he soon became disillusioned by the dictators in Athens and abandoned politics as a career. Later, when the dictators were deposed, he reconsidered politics after a democratic government was established. About that same time, Socrates, the famous philosopher, befriended him and became his mentor. When Socrates was tried and convicted on vague charges and subsequently sentenced to be executed, Plato once again felt repulsion for a democratic government. According to him aristocracy of merit, not wealth, was the best form of government. He advocated equal treatment and consideration of women for inclusion in the guardian class. His political philosophy, while never chosen by any civil government, had a definite similarity to the hierarchy of the Catholic Church. Plato's writings echoed God's ideals that ethical theories be based on the belief that all people desire happiness and that everyone desires to be virtuous. He wrote that since everyone knows that moral virtue leads to happiness they would lead a virtuous life. Plato wrote thirty-six books on the subject of political and ethical questions in a dialogue form that recorded conversations between two or more persons. Many of these dialogues

---

14 "Plato," <u>World book encyclopedia</u>, Vol. 15, p. 502

featured his friend, Socrates. In 387 B.C. he founded the Academy in Athens, considered by many to be the first university which taught philosophy, ethics, justice, and piety primarily but also astronomy and biology. Plato's writings were a major influence on the world, and upon the development of Christianity after his death.

ARISTOTLE[15]

Aristotle, the next "giant," lived from 384 B.C. to 322 B.C. and was Plato's most famous pupil. He became the greatest and most influential philosopher of the times and wrote at least 170 books, of which 47 still exist. Aristotle's writings cover a wide range of subjects and constitute a virtual encyclopedia of the times. He was so admired that his writings, contrary to what he likely intended or wanted, actually inhibited original thought. Everyone believed that if Aristotle wrote it, there could be no other point of view. Some of his teachings, such as "Poverty is the parent of revolution and crime," and "The fate of empires depends on the education of youth," undoubtedly pleased God because they are still considered to be the basis for perfect government. After teaching at Plato's "Academy," Aristotle tutored young Alexander of Macedon. When Alexander the Great became king, Aristotle returned to Athens and established his own academy, the "Lyceum." Because of his father's medical profession, Aristotle wrote extensively about biology, but he also wrote on ethics, logic, physics, astronomy, and meteorology. Aristotle's influence on Alexander became the impetus for new studies and facilities for learning for hundreds of years to come after Alexander built the

---

15 Michael H. Hart, The 100, p. 70

greatest library in the world that contained thousands of volumes. Aristotle's belief and teachings of the earth being fixed and being the center of the universe, however, retarded true understanding of the workings of the universe because everyone had considered him to be the ultimate authority on anything he had written, and thus, if any other theory contradicted Aristotle's, it had to be false. Aristotle's influence on Alexander, and the creation of the basis for future development of information, was very important to Isaac Newton.

EUCLID[16]

Euclid, the best known mathematician of his time, was born in 365 B.C. and lived for 65 years. His most famous work is his writing on geometry, "The Elements." It is a compilation of geometrical knowledge that became the basis of mathematical teaching for the next 2000 years. "The Elements" consists of 13 books and covers plane geometry, number theory, the theory of irrational numbers, and solid geometry.[17] His remarkable compilation of axioms and theorems for the time is one of the significant steps in the intellectual development of God's children. Isaac Newton, as we will see later, absorbed the teachings of Euclid by skimming through his writings. Dr. Isaac Barrow, the first holder of the Lucasian chair of mathematics and Newton's mentor, examined Newton for a scholarship and when he learned that Newton had not studied Euclid, he did not examine him on Descartes because he believed that as he had not studied Euclid, there was no possible way he could

16 Ibid, p.75

17 "Euclid," <u>World Book Encyclopedia</u> Vol. 6 P. 303

understand Descartes. Although Newton did not tell him at the time, he had mastered Descartes.

ALEXANDER THE GREAT[18]

Although Alexander the Great, the greatest warrior of this period, only lived to be 33 years old, he is one of the "Giants." He studied under Aristotle and through his indirect contribution to intellectual growth, he achieved a great deal by the time he died in 323 B.C. He founded more than 20 cities, the most famous of which is Alexandria in Egypt[19] where the greatest library of the ancient world held a magnificent collection of more than 700,000 books. This library attracted scholars and men of letters from many other countries.

ARCHIMEDES[20]

Archimedes, a great mathematician and inventor, lived in Syracuse, Sicily, during the period 287 to 212 B.C. Some of his inventions, the catapult, the compound pulley, and a screw type pump are still in use today in much the same form. His greatest contribution, however, was in geometry. He determined the value of pi (ð) and also proved that the volume of a sphere is two-thirds the volume of a circumscribed cylinder. He considered this his most significant achievement and requested that a representation of a cylinder circumscribing a sphere be inscribed on his tombstone. Archimedes also discovered other fundamental theorems relating to the

---

18 Michael H. Hart, The 100, p. 174
19 "Libraries of Papyrus," World Book Encyclopedia, Vol. 12, p. 228g
20 Daniel Crystal, Biographical Encyclopedia, p. 40

center of gravity of both plane figures as well as solids. But Archimedes' profound concentration with his work ended his life; while intently working on a problem he failed to notice the Romans enter the city. When a Roman soldier demanded that he accompany him he refused to leave until he had finished the problem. The soldier is said to have drawn his sword and killed him. Archimedes' achievements were basic building blocks on which science developed. His methods suggest that, had he lived longer, he may have perfected integral calculus 2000 years before Isaac Newton did.

ERATOSTHENES21

The achievements of the Greek mathematician, Eratosthenes, are still remembered with great reverence today. This "giant," born in 276 B.C. in Syene, a Greek city now known as Shahhat in Libya, lived a long life and died in 197 B.C. He first studied in Athens and then in Alexandria where he became the director of the great library that provided him with the means and the time to study and access to the most complete collection of reference material of the times. His mathematical achievements include a method for determining prime numbers, that is, numbers that can only be divided by 1 and themselves and is still known as the "Sieve of Eratosthenes." Although this process is now achieved by computers, the computer process itself is still very similar to Eratosthenes' "sieve" system. The achievement most often associated with Eratosthenes, however, is his determination of the size of the earth. His studies convinced him that the earth is round and he determined this by observation that at noon on the day of the summer solstice in Alexandria, a vertical post casts no shadow.

---

21 "Eratosthenes", <u>World Book Encyclopedia</u> Vol 6, p.269

He also determined that at the same time in Syene, 500 kilometers to the south, the post did cast a shadow. There is a variation on this story which substitutes a well for the posts, but the principals used and the method of measuring are common to both. Eratosthenes used geometry as developed by Euclid to measure the angle of the shadow. Using the distance between the two towns, he then related the angle measured to that distance and the center of the earth and the number of times the angle is contained in 360°. From that he ascertained the circumference of the earth and translated it to determine the diameter to be 7850 miles (an error of only 50 miles), an astonishingly accurate calculation considering the rudimentary basis used to compute the angles and distances.[22]

Perhaps God watched with great anticipation while mankind accrued all this scientific development and hoped that it would lead to the peaceful existence of mankind. But that was not to be. There had to be a way, however. Perhaps after a good deal of thought and consideration, He decided that the people needed someone stronger and with more influence than the philosophers who had gone before to teach ethics, love, justice, tolerance, and a moral code that would create an atmosphere of peace, love, trust, integrity, and honesty. Jesus was His solution to the problem and He made the momentous decision to put Jesus on this earth to teach the people how to live together.

JESUS CHRIST

Jesus Christ[23] should definitely be considered among the giants because his teachings form the basis for the moral values and principals of Isaac

---

22 David Crystal, <u>Biographical Encyclopedia</u> p.313
23 Michael H. Hart, <u>The 100</u>, p.17

Newton. Newton originally enrolled at Cambridge to study religion but after a few months

He changed his course of study to mathematics and physics. He continued, however, to study religion. He became a religious scholar of renown on his own and although he did not believe in the Trinity, he believed in God and acknowledged Jesus Christ as an individual but rejected the concept of Jesus being divine. Newton studied and accepted the teachings of Jesus as the platform for his personal conduct and as the standard for his relationship with God and with other people. To understand the teachings of Jesus we must draw from the writings of those who knew him. Unfortunately, he did not write anything himself and this causes great concern regarding the proof of his existence. The principal sources of information concerning the life of Jesus are the gospels, written during the 1st Century by those who knew Jesus personally. The lack of additional source material, and the theological nature of the biblical records, caused some 19th century biblical scholars to doubt his existence; however, current day scholars generally agree that Jesus is an historical figure whose existence is authenticated both by Christian writers and by several Roman and Jewish historians. The gospels of Matthew and Luke are consistent in their descriptions of the early childhood of Jesus. Luke also relates the story of the journey of Joseph and Mary with the young Jesus to the temple for the Passover feast. But none of the gospels relate anything of the life of Jesus from the time he was 12 years old until he began his public ministry 18 years later at about the age of 30. The literary relationship of the Synoptic Gospels, the first three gospels of Mathew, Mark, and Luke, agree in their interpretation of the events of Jesus' life and indicate that the public ministry of Jesus began after the imprisonment of John the Baptist and lasted for about one year. The gospel, according to John, indicates that it began with the selection of the first disciples and lasted for about three years. The accounts of the public ministry, and the immediately preceding events, are generally consistent in the Synoptic Gospels. Each describes the baptism of Jesus in the Jordan River by John the Baptist and each records that after the baptism Jesus retired to the nearby wilderness for a forty day period of fasting and meditation. All three synoptists mention that during this period the devil, Satan, tried to tempt Jesus. Afterwards, he returned to Galilee, visited his home in Nazareth, then moved to Capernaum to begin his teaching. It was at this time that Jesus selected the first of his twelve disciples. Using Capernaum as his base, Jesus and his disciples traveled to neighboring towns proclaiming the coming of the kingdom of God. Many Hebrew prophets before him had also done this, but when the sick asked for his help, he sought to heal them by divine power and they witnessed the miracles of Jesus. He taught the infinite love of God for the humble and the weak and he promised pardon and eternal life in heaven to the most hardened sinners, provided that their repentance was sincere. The essence of these teachings is recorded by Matthew in the Sermon on the Mount containing the Beatitudes and the Lord's Prayer. These, then, according to the teachings of Jesus, are the principals by which God desired the people of the world to live and pray for guidance. Although the synoptists generally agree that Jesus spent most of his time in Galilee, John indicates that most of his public ministry was conducted in Judea and records that Jesus made many visits to Jerusalem. His teachings and the miracles he performed at this time, particularly the raising of Lazarus in Bethany, convinced many people to believe in him. The most significant event in Jesus' public ministry, however, was Simon Peter's realization at Caesarea Philippi that Jesus was the Christ although, according to the synoptic gospels, Jesus had not previously revealed this to Peter or to the other disciples. Upon this revelation, and his prediction of his death and resurrection, a voice from heaven was heard that proclaimed Jesus to be the son of God, and thus confirmed the revelation, and formed the primary premise for the claims and historical work of the Christian church.

The life and teachings of Jesus are often matters for dispute and varying interpretation in Christian history, but Jesus and his teachings have survived the test of time and he is recognized as the son of God by Christians the world over. God certainly put him here to inspire the people of the world to live better and more peaceful lives through the Beatitudes and the Lord's Prayer. In the last days before his death, I believe Isaac Newton also considered himself to be a son of God as was Jesus. Jesus certainly inspired Newton and the Keynes manuscripts record his appreciation for the early religion of Jesus Christ. He wrote that it was "the perfect religion prior to the Roman adoption of Christianity as their official religion and the appointment of the pope."

PTOLEMY

The earliest of the giants born after the birth of Christ was Ptolemy,[24] a Greek born astronomer and mathematician who lived in Egypt. It is interesting to note that although he lived between the years of 100 and 170 A.D. his intellectual influence on the astronomical understanding of the universe lasted for nearly 1500 years. During this time a long blank period existed in the history of the development of the technology understanding of how the sun, moon, stars, and planets operated in our system. His astronomical theories dominated scientific understanding until the 16th century. Ptolemy proposed an elaborate mathematical theory to prove his geocentric theory of the workings of the universe that outlined a system whereby the earth was immovable and was circled, in order, by the moon, Mercury, Venus, the Sun, Mars, Jupiter, and Saturn. He believed that the earth was stationary and that the planets and stars moved in perfectly circular orbits. In order to account for the inconsistencies presented by the

24 "Ptolemy", <u>World Book Encyclopedia</u>, Vol. 15, p.754.

theory, he elaborated to explain the backwards motions of the planets and the apparent variations in the size and brightness of the moon and planets. He proposed that the planets, the sun, and the moon moved in smaller circles around much larger circles in which the earth was centered. That made his system fit the observations recorded by other astronomers. The Almagest, meaning Great Work, is a set of 13 books Ptolemy wrote to express his theories and is his most famous work. In other books, he greatly advanced the study of trigonometry. One of the most significant books Ptolemy wrote in terms of historical importance is his Geography. This book charted the world with a system of latitude and longitude and used the Canary Islands as the prime meridian. The determination of longitude is a critical problem for sailors all over the world. A great many of them lost their lives in accidents and collisions which occurred because, with the obvious disadvantages of fog and darkness, captains of sailing vessels were not able to determine their position during a voyage and often collided with mountains or simply ran aground. This problem is discussed in greater detail later but at this point it is interesting to note that this was recognized as a serious problem and one which seemed to defy solution. Ptolemy's system influenced world mapmakers for hundreds of years and is thought to be the inspiration for Christopher Columbus to attempt his voyage to America. Today his system remains the model of all systematic atlases. Although Ptolemy has been accused of capitalizing on the astronomical work of Eratosthenes to determine the diameter of the earth, it is interesting to note that at this early time in the history of the world the earth being round was a well known and accepted fact. Ptolemy's system of longitude suffered from a lack of reliable information: he under estimated the distance from the Canaries to China which he calculated as 180° instead of 130°. However, given the level of information available at the time, it was a remarkable achievement. Ptolemy also calculated a value for pi, even though the Chinese had used 3 since 1650 B.C. When decimals came into use in the 1600s the decimal value for pi was finally determined with more accuracy. The Ptolemaic survived and dominated scientific thinking until the 16th century[25], but there is a huge blank space in the development of science in the 1500 years between the life of Ptolemy and the publication of the theories of Copernicus.

MUHAMMAD[26]

In the 1500 year blank spot in scientific development, God clearly continued in His efforts to create a peaceful world. Apparently He decided that since Buddha, Confucius, and Jesus had not really achieved peace in the world, He would delegate yet another sector of territory to yet another religious leader–Muhammad, who was born in Mecca in 570. Although most Arabs were pagans, there were a number of Jews and Christians in Mecca. When Muhammad, at the age of 40, learned from them of a single omnipotent God who ruled the entire universe he became convinced that this one true God was Allah who spoke to him through the Archangel Gabriel and chose him to spread the true faith. Muhammad was responsible for both the theology of Islam as well as its main ethical morals and principals. In addition, he wrote the "Koran", a collection of Muhammad's statements that he believed had been divinely inspired. The Koran is to the Muslims what the Bible is to Christians and the influence of Muhammad is enormous. But in spite of his influence, Muhammad has not achieved the ultimate goal of world peace that God desires. Muslims, however, are included along with Jews and Christians in the group who worship a single God.

---

25 David Crystal, Biographical Encyclopedia, p.769.
26 Michael H. Hart, The 100, p.3.

NICOLAUS COPERNICUS

What is not popularly known about Copernicus is that he was a priest, a Polish canon, who suppressed his findings for over thirty years, and, because he feared religious persecution, did not allow his book to be published until he was on his deathbed.[27] Copernicus was born in 1473 in what is now Torun, Poland. He was educated in the finest universities and studied medicine and law in Italy where he lived in the home of mathematics professor Domenico Maria de Novara. He studied mathematical sciences which at that time were considered relevant to medicine because physicians made use of astrology in their practices. Maria de Novara influenced geographical and astronomical interests in young Copernicus and since de Novara was a severe critic of the authenticity and accuracy of Ptolemy's Geography, he passed those doubts on to Copernicus. While he lived in Italy he wrote the "Copernican Theory" in about 1513, which explained that the Sun, not the Earth, is the center of the universe. Because he delayed the publication for so long, he was presented a copy of the printed book just before he died of a cerebral hemorrhage in 1543.[28]

GIORDANO BRUNO

Giordano Bruno, a religious rebel teaching at Padua University in 1592, was seized by the Catholic Inquisition after lecturing for only three months, at the same university at which Galileo later taught for more than eighteen years, and was turned over to Rome on capital charges of heresy.[29] During his imprisonment in chains in the Castel Sant' Angelo in

27 Michael White, <u>The Last Sorcerer</u>, p.72
28 Michael H. Hart, <u>The 100</u>, p.100.
29 James Reston, Jr. <u>Galileo, A Life</u>, p.43.

Rome, the Inquisition questioned him ruthlessly and continuously. He had arrived at a radical perception about the cosmos: the Sun, not the Earth, was the center of the universe, and beyond that, he said there were many universes made up of stars and galaxies too infinite to count but all of which were unlimited in space and time. He believed that Christianity was more majestic with his views than with the orthodox Earth centered view, since the infinity of God pointed to the infinity of the heavens. This prophetic vision came from Bruno's soul rather than from any scientific revelation. After eight years of such torture, the Inquisition decided that Bruno would never recant and recommended that punishment be carried out. The Pope approved the punishment but said that there should be no bloodshed. They then inserted an iron spike in his palate and clamped his jaw shut with an iron gag. On February 19th, 1600, several hooded monks, known ironically as the Company of Mercy and Pity, went to the Nona Tower, the secular prison across the Tiber River from the holy prison of Castel Sant' Angelo, and took charge of Bruno. They placed him in a small wagon and started off for the town square. During the slow ride over the stone street, Jesuit and Dominican priests chanted their final appeals to him to recant his sinful beliefs against the scripture and invited him to express contrition for his sins to avoid becoming lost forever in eternal damnation as a heretic. They even offered him icons to kiss but with his iron gag and pallet spike, Bruno had nothing to say. He must have been disappointed that he could not reiterate the words he had spoken to Cardinal Bellarmine, the greatest intellectual in the Catholic church, "I neither ought to recant nor will I. I have nothing to recant, nor do I know what I should recant." Bruno had nothing but contempt for his judges and as they prepared to deliver their sentence, he looked them straight in the eye and said, "In pronouncing my sentence, your fear is greater than mine in hearing it." A crowd gathered as the procession continued its way through the streets and when they reached the public square the crowd had grown very large. As they arrived, the throng quivered with anticipation. The sentence was vague and encoded that on a "Day of Justice" the prisoner should be punished "with as great a clemency as possible and without the effusion of blood." While the priests mumbled prayers about deliverance and charity, they stripped him naked and pressed a crucifix against his face. As he turned away in disgust, excited shrieks rang through the crowd. Giordano Bruno was then burned, bloodlessly, as the priests chanted their litanies.[30]

JOHANNES KEPLER

Only four years after Giordano Bruno was seized by the Inquisition, Johannes Kepler published his "Mysterium Cosmographicum"[31] which argued in support of the Copernican theory and provided a mathematical explanation of the system in which it operated. Kepler was born in Weil, Germany, a small town near Stuttgart and graduated from the University of Tubingen. After accepting a teaching position at a school in Graz, Austria, he elected to leave rather than accept a mandatory conversion to Roman Catholicism. While he was seeking a new position, he formed an association with Tycho Brahe,[32] a great astronomical observer before the invention of the telescope. Kepler joined Brahe as his new assistant and the relationship affected the rest of his life. After Brahe died, Kepler realized that a circular orbit of the planets would not fit all of Brahe's observations. This realization brought him to his most significant discovery; that only an elliptical orbit would agree with and fit all of Brahe's observations. This discovery literally destroyed a belief, common to astronomers for two thousand years, that all the orbits were circular.

---

30 Ibid, p.59.

31 J. V. Field, <u>Biography of Johannes Kepler</u>, p.1.

32 Michael H. Hart, <u>The 100</u>, p.273

Kepler was the first to openly support the Copernican theory while he developed three laws to support the heliocentric theory; first that the orbits were elliptical; second that the line joining the sun and a planet sweeps equal areas in an equal amount of time. His third law described in mathematical terms the relationship between the period of revolution of a planet and the axis of its ellipse. These proven laws of planetary motion were an extremely important base of astronomical understanding for Isaac Newton.

Since Kepler was a contemporary of Galileo, the laws of planetary motion also provided Galileo with a supporting base for his belief in the Copernican theory which later helped him to recognize and comprehend the movements of the moons of Jupiter.

GALILEO GALILEI

Perhaps the greatest of the giants was Galileo Galilei. No doubt Galileo pleased God as his intellect developed during his early years. His religion significantly influenced his unlimited potential for scientific discovery, but Galileo tore away the shroud of mystery from one after another of the puzzles of the world and gave us a whole new look at the workings of the universe. For 1500 years, between Ptolemy and Copernicus, the world had suffered a drought in scientific development. Now was the prime opportunity to develop the theories necessary to discover how the universe worked. His intellectual ability knew no bounds. Galileo proved to be a prolific inventor whose discoveries were more profound than anything the world had ever seen. But the world, or more accurately the Pope, was just not ready for such profound theories and discoveries. Since Galileo lived so recently, and because there is so much written about him, we know a great deal more about the details of his life than any of the other giants. He was arrogant and mischievous, carried a chip on his shoulder, and apparently the details of his childhood would give any parent of a troubled teenager great hope. In his early adult years, he frolicked with a street wench who eventually bore him three children.

Galileo, the oldest of seven children, was born in Pisa, Italy on February 15, 1564.[33] His father, Vincenzo Galilei, was a musician of reknown who was the first to combine poetry and music into the form we now know as opera. His mother, Giulia Ammananti, a bitter and shrill woman from a higher social status than her husband, never let him forget their social differences. There were constant arguments and never enough money. This unhappy woman pestered him constantly to do more work and less music. At the early age of 10 years Galileo was determined never to live his life this way. Four years after he entered the abbey of Vallombrosa for his formal education in the humanities he announced his desire to become a monk. Galileo's father, appalled at his son's declaration, removed him from the school and enrolled him at the University of Pisa where he began studies in medicine. His study of medicine declined as his interest in mathematics increased. He became very outspoken to the point of being accused of disrespect and that, combined with his absence from mandatory lectures, caused the university to notify his parents of the possibility of flunking out of school.

Galileo appeared to have a natural understanding of the phenomena of the universe and always eagerly expressed his opinions. The emphasis on outdated classical authors, such as Aristotle whose teachings were still considered the only truth some 1800 years after his death, annoyed him. When hail fell on Pisa, Aristotle had taught that the larger pellets fell faster than the smaller ones because he believed that the heavier objects accelerated more quickly than the lighter ones; therefore the smaller ones must

---

33 James Reston, Jr, <u>Galileo, A Life</u>, p.7

have originated from a lower heaven. Galileo guffawed at this theory and challenged his professors with its absurdity.

He instinctively realized that mathematics held the secret to his interests and arranged to meet with Otilio Ricci, the official mathematician of the Tuscan court. His obvious fascination with the subject material, coupled with his inquisitiveness and his outspokenness, captivated Ricci. Perhaps the fact that Ricci was such a progressive thinker impressed Galileo. Ricci believed mathematics to be a way of looking at the world and would replace Aristotelian logic in explaining the stars and the planets.

All his argumentive tendencies and personality traits did not endear Galileo to his regular professors; they interpreted his conduct as disrespectful. They tagged him with the unflattering label of "the wrangler" as if he delighted in arguing about anything simply for the fun of it and perhaps secretly hoping to make the distinguished professors look ridiculous.

But he remained humble in regard to the churches in Pisa, especially the cathedral. While attending services one day his mind drifted to an oil lamp which hung by a long wire from a high ceiling. It oscillated steadily back and forth with such regularity that it was like the beating of a pulse. At the age of 19 years this inspired Galileo to rush home and begin experimenting with different lengths of string and different weights until he arrived at a rudimentary device that, simply by varying the length of the string, could measure the rate and variation of a patient's pulse. This ingenious discovery put Galileo back in the good graces of his professors, but only temporarily. The professors who congratulated Galileo promptly stole his idea and, with some additional development, designed a doctor's tool called a "pulsilogia". Credit for this invention went to the distinguished medical faculty of Pisa and the tool remained in use for many years.

This was the first of the Galileo legends. Wherever the truth of the story lies, it describes the central theme of Galileo's life: an act of inspiration born in a religious setting led to an abstract principal that could be tested by experiment and adapted to yield an invention for the lasting benefit of mankind. While Galileo invented devices he hoped would make him rich—such as the pendulum device to measure pulse, the thermometer, the compass, and another pendulum device to raise water—he did not question the church sanctioned myth that the earth was the center of the universe and that a great winch system cranked the stars around the earth every twenty four hours. By 1600 at 36 years of age, he had not yet thought about astronomy.

Giordano Bruno was a professor at Padua University with Galileo, but only three months after he began his lectures he was seized by the Inquisition and turned over to Rome on a capital charge of heresy. Galileo had not thought a great deal about astronomy at that time and was still espousing the Polemic theory, which had the earth stationary at the center of the universe with all other planets revolving around it. Bruno had leapt to a radical intuition that put the sun, not the earth, at the center of the universe. He even went so far as to say that beyond that, there were many universes made up of stars and galaxies too infinite to count. Giordano Bruno, however, had thought a great deal about astronomy and had suffered the terrible consequences of the Inquisition as a result of his heliocentric theories that he delivered to the university students during his lectures. The fate of Bruno did not, however, appear lost on Galileo. The authority of the church and the severity of the punishment for heresy impressed him. But it did not occur to him that this was a pattern of events to come.

Although Galileo is generally credited with the invention of the telescope, Hans Lipershey, a Dutch optician in Middleburg, Holland actually invented it in 1608. Although Lippershey had applied for a patent, for some reason it had not been granted. Nine months later when Galileo found out about it he realized immediately how it must work and what a valuable tool it would be for the military authorities. Before he hardly had a chance to breathe, he learned that Lippershey was on his way to Venice to present his invention to the Venetian state. Galileo realized that he needed to move quickly. He contacted his friend, Friar Sarpi, the scientific adviser to the government, who arranged to block any access to Venetian authorities by foreigners until he could test his own device. He then set about building one on his own and with the typical ability that was his best asset he quickly developed a ten-power scope. Thus his telescope was not only a scientific achievement, but also a political triumph. Eventually he improved its power as well as the field of vision and this fixed his attention to the heavens—and it eventually led him to the path of Giordano Bruno. By 1610 he had outfitted his workshop to grind his own lenses and quickly became the first man to see the mountains and craters of the moon and the stars of the Milky Way. He discovered the rings of Saturn and, perhaps his most important discovery, the moons of Jupiter.

Although Galileo is generally credited with the discovery of the moons of Jupiter in 1610, it was Simon Marius, a German astronomer who was a student of Tycho Brahe, who first discovered the moons a year earlier in 1609, and it was he who named them Io, Europa, Ganymede, and Callisto. Other astronomers, however, did not recognize his claim due to a lack of documentation, and Galileo's claim in the following year was much more credible, thus he gets the credit for the discovery. Galileo, trying to curry favor with the Duke of Tuscany at the time, named the moons as a group and called them the Medicean planets, to honor the family name of the Duke, but recently Marius' names for the moons have become the generally used identification for them.

As he continued to record the changing positions of the moons, his confusion turned to amazement when he realized that the moons were revolving around the planet as the planet itself moved. The similarity of the Earth and its own moon was immediately obvious. With this single discovery had had defeated the principal objection to the Copernican theory and proved the Aristotelian theory impossible; the stars were not fixed and rotating around a stationary Earth. This discovery showed that, contrary to the old beliefs, there were two centers of movement in the universe and not everything revolved around the Earth. Of all the things he had perceived as the first human observer, the moons of Jupiter were the most important—and it was also the most iconoclastic.

He had to publish his findings quickly in order to establish this discovery as his own. He immediately started his manuscript on the "Sidereus Nuncius", or the "Starry Messenger". By March 1610, five hundred fifty copies of the Starry Messenger were distributed to the most prominent people and it instantly became the most important book of the 17th century at that time. It forever changed the way we view the universe, and it also set into motion a chain of events that had serious consequences for Galileo Galilei.

DESCARTES

Rene Descartes[34], born in France in 1596, was also a contemporary of Galileo. He is the father of analytic geometry which interested Isaac Newton much more than Euclid's plane geometry. But Descartes also pioneered the attempt to formulate simple universal laws of motion that govern all physical change. He was educated at a Jesuit college and lived in the Netherlands where he produced most of his scholarly writings. Descartes, like Newton, developed rules for reasoning and firmly believed in God. He died in 1560, only a few years after Galileo.

Obviously some giants influenced Newton more than others, but it is also obvious that Isaac Newton was the beneficiary of a great deal of vision and hard work by his predecessors. It is truly amazing to consider all the things these individuals have achieved, particularly in light of the state of the art at that time. It is also difficult to believe that so many things, such

---

34 Michael H. Hart, <u>The 100</u>, p.248.

as a method to determine longitude at sea, escaped a solution until so recently as the late 18th century.

But none of this can distract from the magnificent achievements of Isaac Newton. It taxes the imagination to understand how he discovered and invented so many things at such an early age without the aid of any reference material at all. Original discoveries that are still used today and which enhanced the productivity of subsequent inventors put Newton in a class by himself as the greatest human being of our planet. As we shall see in subsequent chapters, Newton's ability to achieve was most certainly enhanced by divine guidance.

Chapter 3

Galileo's Dilemma

In the 16th and 17th centuries the Catholic Church was the guardian and absolute authority over matters of church doctrine, and the state, particularly Italy, acquiesced to the power of the Pope. Heretics were the enemy of the church and anyone who espoused anything which contradicted church doctrine was considered a heretic. The Pope and his Cardinals kept a close watch on any development which might threaten the position of the church, and the Pope reserved the exclusive right to defend the faith and unilaterally judged any attack on its doctrine.

After Galileo discovered the moons of Jupiter he realized that the Copernican theory of the universe was proven by the orbiting moons beyond any doubt. Although the fate of Giordano Bruno was fresh in his mind, it did not prevent him from voicing his opinion that the sun, not the earth, was the center around which all the other planets revolved. The hierarchy of the Catholic Church than began to regard Galileo with new interest.

Over the years Galileo challenged the authority of his superiors and irritated many of the influential leaders in the scientific and religious communities. His abrasive and dogmatic attitude alienated some of the people in positions of authority whose help and cooperation he now needed. As challenges and barbs were passed back and forth between Galileo and members of the Roman Catholic Church, the contest became more concentrated in individuals and Father Horatio Grassi, who held the chair for mathematics at the famous Collegio Romano, emerged as the primary Catholic adversary. Father Grassi was also an excellent scholar, a fine writer, and a experienced infighter, Grassi's writings about astronomical matters were gaining wide circulation in Rome and were being used to discredit the Copernican view of the universe.[35] If Galileo had been able to restrain his sharp tongue, he and Father Grassi might have had some opportunity to communicate in a useful dialogue, but Galileo could not restrain himself and, in the word of James Reston, provoked a mongoose.[36]

Father Grassi minced no words in his challenge for "all who are dutiful will call everyone away from Copernicus and will reject and spurn his hypotheses." With this, Father Grassi and the Jesuits had thrown down the gauntlet and it was now for Galileo to decide whether he dare to pick it up.[37] There were three events in the ensuing months which influenced Galileo's decision. First, the Grand Duke of Tuscany, who had been Galileo's friend and mentor and for whose family he had named the moons of Jupiter, died after several years of ill health. But as he lost a supporter, he also lost an antagonist; Cardinal Bellarmine, who had been Giordano Bruno's prosecutor in the Inquisition, died also. Bellarmine had also warned Galileo in a letter of the danger and folly of publicly espousing the Copernican theory. In addition, Galileo received a cordial letter from Cardinal Maffeo Barberni thanking him for the help he had given his cousin, Francesco Barberini, in attaining a doctorate at the University of Pisa. Cardinal Barberini also declared his affection for Galileo and the affection of his house, and went on to say "We are ready to serve you always". Only two weeks later, Pope Gregory died and Cardinal Barberini was elected Pope. He took the name Urban VIII.[38]

35 James Reston, Jr. Galileo, A Life,p.179
36 Ibid, p.180.
37 Ibid, p.181.
38 Ibid, p.187

Galileo was convinced by the turn of these three events that he could now undertake an attack on the Jesuits, believing that the absence of the chief prosecutor of Bruno and the elevation of his friend to Pope, would combine to provide him with the necessary influence to gain the tolerance of the church for the intellectuals and scientists who espoused the Copernican theory of the universe. He was wrong.

There were other members of the church who shared the opinions of Bellarmine and Grassi and who pressed their arguments forcefully. The accusers were armed with such selective portions of doctrine as Psalm 93: "The Lord reigneth. He is clothed with majesty and with strength; the world is also established so that it cannot be moved."[39] Another priest charged that "Galileo's pretended discovery vitiates the whole Christian plan of salvation." Another declared that "it cast suspicion on the doctrine of the incarnation". And another reasoned that if the Earth was a planet like other planets, the others must also be inhabited since "God makes nothing in vain". How can their inhabitants be descended from Adam? How can they trace their origin to Noah's ark? How can they have been redeemed by the Savior?"

All these words drifted into the Vatican and on May 17, 1611, the Congregation of the Holy Office—the Inquisition—took up the case of Galileo and his Starry Messenger described in the previous chapter to decide whether those writings were heretical.

In letters to Father Castilli, Galileo argued forcefully that "What God would give man language, intellect, and senses and then expect him to bypass their use?" The letters received wide dissemination and pleased Galileo who believed that surely reason would win out in this argument.

---

39 Ibid. p.120

But that was not the case. The letters alerted churchmen, both humble and powerful, of the arrogance of an impertinent layman. Galileo had dared to trespass on the realm of theology—the queen of all sciences. H e had been bold enough to interpret scripture, to question its reason, and to take upon himself—a layman—the task of defining the Holy Scripture. This was, by definition, heresy. Defining Holy Scripture was the sole and exclusive province of the Pope.

On April 12, 1633 the Inquisition summoned Galileo to the Vatican for interrogation.[40] At this point in time the Pope ruled as supreme authority in these matters and the state remained aloof regarding matters of religion. Galileo's refusal to appear before the Pope would result in his immediate excommunication from the church and he would be lost forever in eternal damnation as a heretic. Galileo loved and respected his church and was faced with a fateful decision whether to respond to the summons of the Inquisition or face the excommunication of the church. He decided, finally, that he must face his accusers and defend himself and his intellectual position on the Copernican theory.

He was continually questioned during his two month imprisonment in the Vatican. Finally the Inquisition met with the Pope on June 16, 1633, and they discussed torture. He was to recant his beliefs before a full session of the Congregation of the Holy Office, to be imprisoned for life, and in the future not be permitted to utter a single word about cosmology or to discuss the Copernican theory—or even teach its opposite: the Ptolemaic theory that the earth was fixed in a stationary position and was the center of the universe. On June 22, at almost 70 years of age, Galileo, a weary

---

40 Ibid, p.246

and broken old man, retracted his beliefs as he had been ordered to do and his sentence was formally announced.[41]

Later, upon his appeal in December of that same year, he was permitted to return to his own home with the requirement that he live the strictest control of the local priests, to entertain no one, nor to allow his friends to gather. If he violated any of the rules or become a nuisance, he would be returned to Rome and the prison in the Vatican. He lived in isolation and silence knowing he had been betrayed by his friend, the Pope, and many others of the clergy who had refused to express their secret, personal beliefs in the Copernican theory.[42]

Dava Sobel, in her latest book, *Galileo's Daughter* describes the extent to which the church enforced his sentence. His daughter was in a convent only a very short walk away, but her vows prohibited her from leaving the premises of the convent and his sentence to isolation in his home would not permit him to go the short distance to visit her, so they were reduced to exchanging letters for the remainder of his life. He lived like that for another eight years until his death in January 1642.

Three hundred fifty years later, on October 31, 1992, a panel of Cardinals, earlier appointed by Pope John II, gathered to render their report on the Galileo affair. One of the Cardinals read an acknowledgement that Galileo had been right and that the Earth does turn and revolve around the sun. The Pope then formally received and approved the findings of his commission and turned to his own formal declaration. Without lifting his eyes to his audience he read, without emotion, "The lessons of the Galileo affair remain valid," he said "and could be relevant in the future, especially in the area of biogenetics". There was no apology.

---

41 Ibid, p.260
42 Ibid, p.270

Neither was there any further explanation of the meaning of the Pope's words regarding the relevance of the lessons of the Galileo affair in the area of biogenetics.

What the Roman Catholic Church did to Galileo has been both savored and scorned for nearly 400 years. What the church did to itself, by virtue of its own actions, has been that of equal fascination and avoidance. And what the church did not do until October 31, 1992, to heal this historical wound remains a vibrant, modern question. The face of Galileo still haunts the Catholic Church.

Taking three hundred fifty years to resolve the Galileo matter seemed an extraordinary long period to James Reston, Jr., the author of a magnificent biography on Galileo. He traveled to the Vatican in April 1993, for an audience with Cardinal Poupard and asked why it had taken so long. The Cardinal answered that an "interdisciplinary exercise such as this takes time." He inquired as to whether the Pontiff's statement of October 31 should be considered a formal apology and the Cardinal answered "not at all; it was merely a formal recognition of error." When asked if the church would ever have anything further to say about this case, he replied, "Why? It is done, finito." And when Reston asked if Giordano Bruno would be next, the Cardinal merely smiled.[43]

God was surely very disappointed with the outcome of the Galileo affair and He took immediate steps to correct the terrible wrongs that had been committed. The Inquisition, headed by God's own representative on Earth, had first handled the matter of Giordano Bruno in a very heavy handed and sad manner. Secondly, the matter of Galileo forced re-examination of every personal relationship with the hierarchy of the Catholic

---

43 Ibid, p.285.

Church. Galileo came very close to achieving God's purpose: to determine, for the benefit of mankind, how the universe operates. His God given task failed because of the short sightedness of God's representative on Earth, the Pope. Galileo's discovery of the moons of Jupiter is positive proof that the planets moved and that their moons orbited around them. The fact that the Catholic hierarchy did no accept that proof continues to remain a mystery.

God's solution was to send the world Isaac Newton in that same year, and to make His message clear, had him arrive on Christmas Day and wrapped him in a protective sheath to protect him from any fate similar to Galileo's for all his life.

Chapter 4
The Church of England

At the time of Newton's birth the country had just entered into the Stuart monarchy era and the Reformation had been firmly established by an act of the state. But during most of Isaac Newton's life, England was in a state of turmoil and unrest. There were civil wars, foreign wars, the plague, and a continuous period of religious dispute and disorder.

The Pope refused to annul King Henry VIII's marriage to his first wife, Catherine of Aragon, who had not borne Henry a son. Ann Boleyn, Catherine's maid of honor, although not beautiful, was extremely clever and had the resourcefulness to enchant her royal admirer. The king wanted to marry Ann Boleyn in the hope that the marriage would produce an heir to the throne. The story of Ann Boleyn is well known and reflects the atmosphere and temperament of the monarchy. The Parliament, at the king's insistence, broke with the Catholic Church and established Henry as head of the Church of England. Henry had secretly wed Ann several months earlier and three months after being crowned queen she gave birth to Elizabeth. But Henry soon began to tire of Ann, disappointed because he wanted a son, and three years later condemned her on a charge of unfaithfulness. Ann was beheaded in May 1536. After Henry died, leaving as the only claimants to the crown the descendants of his sisters. Elizabeth became queen and although Henry had favored the children of his younger sister, Mary to succeed him, Elizabeth's councilors chose James, the son of Henry's older sister. James became king in 1603 at a time when experience and wisdom would have been helpful; James had little of the former and none of the latter.

James considered his monarchy a divine right and believed he was accountable only to God.[44] According to his beliefs, his subjects had no rights; perhaps privileges granted by royal decree, but otherwise they had only duties.

Early in his reign James experienced a serious religious problem, the "millenary petition", so called because the thousand Puritan clergymen who signed it asked for reforms in the church. James denied the petition but did approve the request for a new translation of the Bible which became known as the King James Version. He continued disapprove of the puritans, Presbyterians, and Separatists because they opposed royal authority.

Catholics had long opposed the recusancy fines levied upon them for failing to attend services of the Church of England or to recognize its authority. Some angry Catholics plotted to capture the king and force him to accede to their demands, but other Catholics betrayed the plot and out of gratitude James allowed the fines to be halted. When those who stopped attending church grew to alarming numbers, James began to collect the fines again. This vacillating policy caused a group of Catholics to attempt the "Gunpowder Plot". In order to take over the government they planned to blow up the chamber where the king and lords assembled when parliament met in the autumn. Although the coup was discovered and the plotters apprehended, it revived the fear of Catholics that had been in remission since the victory over the Spanish Armada in 1588.[45]

44 W. E. Lunt, <u>History of England</u>, p.388.
45 Ibid, p.390.

The parliament of 1604 stayed in session for seven years debating the relative powers of the king and parliament, This seven-year period began a contest between the king and the parliament that was not settled until the revolution of 1688.

Under James the persecution of Protestant groups, such as the Puritans, caused the groups to migrate to America in 1620 where they formed the Plymouth Colony in Massachusetts. Jamestown, the first permanent English colony in America, was named in honor of James, but his only interest in colonies lay in Northern Ireland where he seized land from Irish Catholics and gave it to English and Scottish Protestants. This caused conflicts and ill feelings that still exist today.

After James died in 1625 his son, Charles, became king. Following in his father's footsteps, he too believed in the divine right of kings to rule. Between 1625 and 1642,Charles dissolved one parliament after another because the members refused to submit to his demands. He continued to rule for eleven more years between 1629 and 1640 without a parliament. Finally, he summoned what became known as the "Long Parliament" which met from 1640 to 1653 and held its last session in 1660. When the king tried to seize five leaders of the parliament in 1642, civil broke out. This is the tumultuous situation into which Isaac Newton was born.

Oliver Cromwell became the leading general of the parliamentary army and the victor in many important battles. Finally Charles fled to Scotland in 1646 and the war ended which propelled Cromwell into the leadership position and he governed as a protectorate. Within a short time Scottish leaders turned Charles over to Parliament and the army seized him. But he soon escaped and made a secret deal with Scotland. A second civil war began in January 1648, when Newton was only six years old, and lasted seven more months. Charles was captured again, convicted of treason by Parliament in 1649, and beheaded.

Charles II, the son of Charles I was restored to the Stuart line and in 1651 he was proclaimed king of Scotland. Meanwhile, Oliver Cromwell continued to rule as a protectorate until he died in 1658. By this time the people had become dissatisfied with the protectorate and invited Charles to return and proclaimed him king in 1660. His reign was marked by two wars with the Dutch from 1652 to 1654 and from 1664 to 1667, by the Catholic Terror, and the plague.

The Catholic Terror was a well-planned plot to dethrone Charles, seize the government and re-establish the Catholic Church as the national religion. The plot failed, but succeeded in fanning the flames of the smoldering fears which had haunted popular feeling for many years and as a result the public erupted into a state of unreasoning panic. Parliament then passed laws requiring its members to express disbelief in several Catholic doctrines and excluded Catholics from election to Parliament.[46]

The plague which struck in the summer of 1665 lasted two full years until the spring of 1667. Cambridge University and all its colleges were closed during this entire period in order to protect the students and stop the spread of the terrible disease. Newton went home and studied without books or a tutor for the entire period of time and it was during this terrible event that he invented calculus, well before he graduated from Trinity College.

During the reign of Charles II, the parliament shared authority with the king, and when Charles died in 1685, his son James, became King James II. He was a Roman Catholic and wanted to restore Catholicism and revert to absolute monarchy. The people did not like James' ideas, but disliked civil war even more. For a short time they believed that James' Protestant daughter would become queen so they reasoned that it would be better to tolerate the Catholic James for a while rather than risk another civil war. When James had a son, they realized that Catholicism would be

---

46 Ibid, p.456

permanent and many people demanded his abdication. Leaders of Parliament then revolted and invited James' Protestant daughter and her husband, William of Orange, the ruler of the Netherlands, to become rulers. So in 1688, William and Mary landed in England and James gave up the throne and went into exile in France.

With all these things going on one after another, and sometimes simultaneously, it is easy to understand the turmoil in Isaac Newton's life. The sequence of events seemed to be never ending. Religious unrest and successive changes in the official religion from Protestant to Catholic and back again, the civil wars, and the plague all combined to threaten life itself as well as make his tenure at Trinity College pretty uncertain.

Chapter 5

The Link Between Science and Religion

Isaac Newton grew up the small hamlet of Woolsthorpe, just north of Cambridge in England, and probably never ventured more than ten miles from his birthplace before he went to college. Because of his premature birth—Newton was so small he could fit into a quart pot—his chances of survival were slim. Newton's father had died several months before Isaac's birth so his mother raised him in the same small town from which his entire family descended. In fact, all the Newtons came from towns and villages within five or ten miles of Woolsthorpe.

Three years after Newton's birth, his mother, Hannah Newton, married Barnabas Smith, a clergyman from a nearby town. But Smith, recently widowed, needed a wife, not a baby son. He declared to Hannah that he did not want the boy so the traumatized little three-year-old Isaac went to live with his unaffectionate grandmother Ayscough.[47]

Reverend Smith had inherited a theological library from his father, who had also been a clergyman, and Isaac Newton, very conscious of religion as

---

47 Frank E. Manuel, <u>A Portrait of Isaac Newton</u>, p.24

a child, somehow acquired nearly 300 of those books. One of the volumes was a large notebook which Barnabas Smith had begun in 1612, some 33 years before his marriage to Hannah Newton, which contained a set of theological headings and only a few notes from his readings. This notebook, a huge reservoir of blank paper, may well have become the birthplace of calculus. It is also believed that the depth and detail of the library established the basis for Newton's theological interest that was well beyond his stepfather's interest or understanding.[48]

Even though many relatives lived close by, there is no record that young Isaac ever had a close relationship with any of them. An uncle, the Reverend William Ayscough, who only lived two miles away, had no contact with him either.

To establish a relative time perspective, the religious upheaval described here and in the previous chapter was the norm rather than the exception in England. It was the period when the Pilgrims escaped religious persecution by fleeing to America, while the crown changed back and forth several times between Protestantism and Catholicism.

In 1653, when Isaac Newton was ten years old, the Reverend Barnabas Smith died and his mother returned to Woolsthorpe. But the joy of having his mother back was diminished considerably by the fact that she brought with her a half brother and two half sisters. Obviously, with three younger children, Isaac did not get her undivided attention. Whatever happiness he realized with his mother's return he lost within two years when he was sent off to grammar school in Grantham.[49]

In addition to studying the books given to him by his stepfather, Newton received religious training at school and perhaps from his uncle

---

48 Richard S. Westfall, <u>Never at Rest</u>, p.51.
49 Ibid, p.55.

who we know was very conscious of his religious obligations. Newton even set about writing a self examination of his conscience and wrote lists of his sins. One list recorded his wish to set his mother's house on fire–with her and his stepfather in it. Another listed such sins as having unclean thoughts, not keeping the Lord's Day as he should, and several other sins of that magnitude.

Because he began his young life with personal experiences that clearly focused on the distinction between a true father and a false father, it enabled him to later distinguish between true gods and false gods. In his later years, he developed into a serious student of religion and demonstrated his ability to quickly perceive the scriptures of the Protestant ministers, the teachers of Jewish religion and law, as well as the Catholic hierarchy, and the medieval commentators. He studied everything about all kinds of religion and ultimately became an extremely well informed theologian in his own right.[50]

Isaac Newton's intense desire to have complete and total order in everything went far beyond religion as we know it today and extended to primitive times and legends of Greek mythology. Not satisfied with the unexplained beginnings of ancient kingdoms, Newton tried to establish a specific and accurate chronology of all the crucial historical events in order to establish absolute dates for such events as the Argonautic Expedition and the fall of Troy.[51] These events can be described briefly as follows:

The Argonautic Expedition is a legend from Greek mythology telling the story of Greek heroes including Castor, Pollux, Hercules, Orpheus, and Telamon, describing their voyage with Jason to capture the Golden Fleece, the golden wool of the flying ram. Their ship, the Argo, was the

---

50 Matt Goldish, Judaism in the Theology of Sir Isaac Newton, p.11
51 Richard S. Westfall, Never at Rest, p.83.

largest ship ever built at that time. After many adventures the Argonauts found the Golden Fleece in the far away land of Kolchis, an area which lies east of the Black Sea. The fleece was guarded by a dragon which never slept, but Medea, the daughter of the king of Kolchis fell in love with Jason and she put a spell on the dragon so the fleece could be captured. There is no date associated with this event.

Troy was a city founded at the southern end of the Dardanelles sometime during the period between 3000 B.C. and 1200 B.C. This ancient story is the basis of poems of the Greek poet, Homer; the Iliad and the Odyssy in which he describes a battle against the Trojans, the inhabitants of Troy, and which used the famous Trojan horse, a large wooden horse inside of which several Greek soldiers hid. Seeing the Greeks depart, the Trojans thought they had given up and believing the horse to be a symbol of good luck, brought it into the city. That night while the Trojans slept, the Greek soldiers came out of the horse and opened the gates to the city to permit the Greek army to complete their conquest. The Trojan horse has since become a synonym for a ruse or trick. Archeologists have discovered as many as seven cities of Troy, each one built on top of another, and have never agreed on which city was in existence at the time of the fable.

Newton never doubted the existence of God. He believed that the perfect arrangement and distribution of planets could not have happened by chance. He also believed that comets were a phenomena in whose progress God had to intervene from time to time. He said that there had been incidents of major cosmic or geological catastrophes in the past and that re-populating the Earth after such an occurrence required a divine decree.

Later in his life he suggested that God guided the comet we know today as "Halley's Comet" to Earth to create the Great Flood. The Keynes Manuscripts at King's College in Cambridge contain a copy of a letter to Thomas Burnet giving Newton's theory of the creation of the Earth and its subsequent development. Newton believed that all the planets and the sun at first had one common chaos and the spirit of God moving on it had separated it into several parcels, and each parcel became a planet. He said that at the same time the sun also separated from the rest of the chaos and began to shine before it formed into its compact and well defined body.[52]

Newton spent over 20 years studying and writing more than one and a half million words exclusively on religion. None were ever printed. There are at least two truly outstanding books on Isaac Newton's religion. One is "The Religion of Isaac Newton" by Frank E. Manuel which points out that in the18th and 19th centuries Newton was cited to illustrate the compatibility of science and faith and asking "if the greatest of all scientists was a believer, how could any ordinary mortal have the impudence to doubt?" Isaac Newton had an especially poignant feeling about the Father who was in heaven, a longing to know Him, to obey and to serve Him. The second truly outstanding work on Newton's religion is "The Judaism in the Theology of Sir Isaac Newton" by Dr. Matt Goldish. Dr. Goldish points out that Newton's pursuit of theological truth began in his student days and continued throughout his life.

Newton's religious beliefs were severely tested when he was forced to decide whether to accept ordination in the Anglican clergy as a condition for retaining his position as the Lucasian Chair of Mathematics at Trinity College. Although he had not come to grips directly with the question of belief in the trinity, he suddenly had to face it squarely in the eye. His honesty and morality would not permit him to deceive himself or the college. Although he did not want to be ordained, neither did he want to give up the position which afforded him access to the scientific world and, of course, provided him a living. Another example of the divine guidance which followed Newton throughout his life occurred when Isaac Barrow, who had held the Lucasian Chair before Newton, used his influence with

---

52 Isaac Newton,<u>Keynes Manuscripts</u>, Kings College

the royal court and arranged for a letter of dispensation. The letter specifically exempted the Lucasian Chair position from the requirement to be
ordained and Newton's name was never mentioned. He never asked for
dispensation nor did the record ever reflect any consideration for him by
name. It just happened without any fanfare at all.[53]

Professor Manuel's magnificent insight into the subject in "The
Religion of Isaac Newton" reveals that Newton saw himself as the interpreter of God's will. Newton believed himself to be the vehicle of God's
eternal truth and that by using his new mathematical notations, calculus,-
the universal language to replace the language lost at the Tower of Babel,-
with experimental methods, he was able to combine the knowledge the
priest-scientist of the earliest nations, of Israel's prophets, and of the Greek
mathematicians. He believed that nothing had been withheld from him.
God had revealed everything to him to demonstrate the solutions to the
riddles of the universe to all mankind. But this blessing had its price. He
was so concerned about his discovery and belief that he assured his friends,
and himself, that he had broken no prohibitions against revealing the
solutions to the riddles of the universe; he had merely expressed it in
another language that the ancient people had known before him.

Newton believed ancient Christianity to be the perfect religion that
became corrupted after being recognized as the official religion of the
Roman Empire by Constantine in 313. Newton believed that the papists
had departed from the unity, they were worshippers of persons as gods, and
adorers of relics to which they interpreted powers they did not possess.[54]

After Newton's death, John Maynard Keynes, famous for his economic
theories, acquired many of Newton's manuscripts and contributed them

---

53 Richard S. Westfall, <u>Never at Rest</u>, p.333
54 Richard Westfall. <u>The Life of Isaac Newton</u>, p.140.

to King's College in Cambridge. Peter Jones, the Librarian at King's College, edited the manuscripts to duplicate them on microfilm to be available for interested scholars. Dr. Goldish, in his "Judaism in the Theology of Isaac Newton" observes that Keynes wrote that Newton was a magician because he looked at the whole universe as a riddle; a secret which could be read by applying pure thought to certain experience–certain mystic clues which God had laid out about the world to allow a sort of philosophic treasure hunt. The key element in Newton's religion is that God revealed a true, simple religion to Adam which consisted of two elements: love of God and love of man. Newton believed that when religion went beyond that, it became corrupt.[55]

Newton knew the Book of Daniel in the Old Testament and the Book of Revelation in the New Testament contained all the clues necessary to solve the riddles of the universe. All one has to do, he said, is study diligently and pray over them. The Tabernacle and Holy Temple of Jerusalem, which contained a pyrtaneum designed as a microcosm to represent the heliocentric universe, fascinated Newton. He believed that ancient scientists had understood the nature of the universe before him.[56]

The sheer volume of Newton's writings on prophecies indicates that he spent a large part of his life working on them–perhaps even more than he did on science. "Search the scriptures by frequent readings and constant meditation on what thou readest", he said, "and pray to God for enlightenment and understanding if you desire to find the truth." Isaac Newton clearly saw himself as the one to whom God gave the wisdom necessary to unravel the divine plan hidden in the prophecies. He stopped short of believing he had the wisdom to predict the future, but he knew God had

---

55 Matt Goldish, <u>Judaism in the Theology of Sir Isaac Newtom</u>, p.4
56 Ibid,p.12

provided him with the ability to solve the mysteries of the universe. Newton had no doubt that since the fullness of knowledge had been revealed to him, his election by God had been empirically demonstrated. He also believed it to be his duty to reveal the solutions to all mankind.[57]

The more frequently Newton's theological, chronological, and mythological work is examined as a whole and placed by the side of his science, the more apparent it becomes that in his moments of grandeur he saw himself in action as the last interpreter of God's will who lived on the eve of the fulfillment of the times. He was the vehicle of God's eternal truth; by using his new mathematical notations with experimental methods he combined the knowledge of the priest–scientist of the earliest nations, of Israel's prophets, of the Greek mathematicians, and of the medieval alchemists. Nothing had been withheld from him. Newton's frequent insistence that he was a part of an ancient tradition, a re-discoverer rather than an innovator, is subject to a variety interpretations. In one of the manuscripts of the Principia that dates from the end of the 17th century, Newton expounds his belief that a whole line of ancient philosophers held to the atomic theory of matter, a conception of the void, the universality of gravitational force, and even the inverse square law. In part, this was an interpretation of myth. Many of the Greek gods and demigods were really scientists. It was the survival of a major factor of the Renaissance tradition of knowledge and its veneration for the wisdom of antiquity. But the doctrine may also take us back to the causes of Newton's profound religious emotions with which we began. The arrogance of discovery terrified him and as if to placate God the Father, he assured his intimates and himself that he had broken no prohibitions against revealing what was hidden in

57 Ibid, p.58.

nature. He had merely uttered, in another language, what the ancients had known before him. To believe that one had penetrated the ultimate secrets of God's universe and to doubt it, to be the Messiah and wonder about one's being anointed, is the fate of the prophets.

Newton's conviction that he was the chosen one of God was accompanied by the terror that if found unworthy he would provoke the wrath of God–his Father. This made one of the greatest geniuses of the world also one of its greatest sufferers. During his lifetime, Newton's extensive writings on religion left nothing unexamined. The sheer volume of Isaac Newton's manuscripts on the prophecies indicates that he spent a very large part of his life working on them. Newton believed he was in direct relationship with his Father and things were revealed to him as they had once been to the Hebrew prophets and the apostles and the legendary scientists of the ancient world whom he identified with one another.[58]

He clearly saw himself as one given the wisdom but not the prophecy, to unravel the divine plan hidden in the prophecies. "Search the scriptures by frequent readings", he said over and over, "and by constant meditation upon what you read and earnest prayer to God to enlighten your understanding if you desire to find the truth." He believed that all the secrets of the universe, including the heliocentric universe itself, were clearly spelled out in the scriptures. He established a set of rules for the interpretation for the Books of Daniel and Revelation.

Several of Newton's followers subscribed to his theories. William Whiston, who succeeded him to the Lucasian Chair, wrote a proof that sustained the proposition that precisely 1700 years after the creation a comet passed the Earth and that its atmosphere and tail caused the great Deluge.

---

58 Frank E. Manuel, <u>A Portrait of Isaac Newton</u>, p.29

In 1802, long after Newton's death, Henri de Saint Simon, a French noble, summoned his contemporaries to found a new church under the hegemony of science-priests and called it Religion of Newton.[59] Another French aristocrat, Champlain de la Blancherie, issued a manifesto roundly denouncing the English nation for its failure to honor Newton's divine person, re-dated the calendar to Newton's birth, and proposed the establishment of a sanctuary at Woolsthorpe.

Isaac Newton, a learned theologian in his own right, wrote greatly detailed manuscripts giving expression to a theology of glory. He wrote in effusive language, a genuine, almost rhapsodic, wonderment at the complex and infinite powers of the Creator.

---

59 Frank E. Manuel, The Religion of Isaac Newton, p.53

Chapter 6

Miracles of Science

There is probably more written about Isaac Newton than any other person in the world. And the number of authors is astounding! It seems as if nearly everyone has written something about his life. It is almost redundant to reiterate the life story of Sir Isaac Newton, but each author has a different perspective of the facts, all in varying degrees from one another. My perspective is unique; I haven't found a single author who has interpreted the facts of his life as I have. It is difficult for me to understand how so many have come so close to my perception, but have stopped short. Perhaps they are reluctant to apply their name to an idea that is off the centerline of our standard religions. Several of my acquaintances have raised an eyebrow at the title of this endeavor as if it is sinful. Could it be just a coincidence that Isaac Newton was born in the same year as Galileo's death? Remember that Galileo was very close to solving the mysteries of the universe when the Pope forced him to recant his beliefs and sentenced him to life in prison. Can it also be mere coincidence that he was born on Christmas Day? Do you believe it a coincidence that Halley's Comet appeared on Christmas Day in 1758, the first time its return was predicted? What protected him from punishment as a heretic when he decided he could not accept ordination in the Anglican Church?

My perception is relatively simple. I believe God put Isaac Newton on the Earth for the specific purpose of solving the riddles of the universe and to tell everyone in the world about it in a universal language that everyone could understand—no matter what language they spoke. The universal language, calculus, was necessary to replace the language lost at the Tower of Babel. Newton's first invention is the standard of mathematics, and is known and used throughout the world.

Even as a young boy, Isaac Newton provided clues to the extraordinary ability that enabled him to understand and unravel the mysteries of the universe. He was an intense thinker who concentrated on a problem until it was solved. Sometimes that took several days during which he forgot to eat or sleep, so intense was his concentration. Concentration and persistence combined with a remarkable level of intelligence, produced the greatest scientist the world has ever known. What is not commonly known is that he dedicated his entire life and work to the glory of God. He believed he had been given the knowledge and wisdom to solve the mysteries of the universe and reveal them to mankind. From adolescence on, his pursuit of knowledge as power over things and his knowledge as a revelation of God were ever present, not as contradictory alternatives, but as directly related with one another; the positive recognition of the relationship between science and religion.

There are numerous stories about him as a lad. When he went to town for supplies, he always carried a book to read. Once, on the way back up a steep hill where it was customary to dismount and lead the horse, his concentration on his book was so intense that he walked right past his house! The horse had slipped its bridle and headed home without him. Isaac continued walking on past his house with his nose in the book and the bridle dragging behind him. As an adult, his concentration caused him to neglect to eat and sleep for several days at a time while he worked out the solution to different problems.[60]

---

60 Richard S. Westfall, <u>Never At Rest</u>, p.64.

Isaac Newton grew up in civil wars. The first, in the year of his birth, lasted four years; when he was six years old, a second civil war broke out that lasted for another six months. Some of the battles were fought within earshot of his home and one of them ended with King Charles being captured, convicted of treason, and beheaded. Before Newton turned 25, there had been two wars with the Dutch, the Great Fire of London, the Catholic Terror, and the plague. And within his lifetime, there was also a rebellion, eleven different rulers and, although the Reformation was firmly in place, the personal religion of the rulers and thus the official religion of England, went from Protestant to Catholic and back again several times. Consequently, the religious leanings of the monarch resulted in governmental changes, which created great anxiety among the people.

The people of England were extremely suspicious of the Pope and the Catholic religion in general. Several threats by Catholic groups against the government and government against the Catholics, such as banning them from election to the parliament, resulted in constant turmoil. It is clearly evident that Newton grew up in very turbulent times. During this period of time many people were fleeing from the religious torment in England and emigrating to America. This greatly concerned Isaac Newton because he also had some disagreement with the religious activity in his native England.

There was never a shortage of religion in Isaac Newton's life, however. Although no one on the Newton side of the family could sign their own name before 1642, his mother, Hannah, was able to read and write and at least one member of the Ayscough family, Hannah's brother William, received his Master of Arts degree from Cambridge and became a clergyman of the Anglican church. Newton's handwritten manuscripts suggest that, since he had no father of his own, Isaac thought of God as his father. During the years he lived with his grandmother, he attended a nearby school in Woolsthorpe. In 1653, when Newton was ten, the Reverend Smith died and his mother returned to Woolsthorpe. Newton spent two years with his mother before attending grammar school in Grantham, about seven miles away. Because of the distance to school, arrangements had to be made for Isaac to live with Mr. Clark, the apothecary. Clark liked him and spent considerable time teaching him how to make things with his hands, which Newton loved to do. Although in this area windmills are fairly rare because there were lots of streams and rivers to supply motive force for water wheels to grind corn, Newton built an exact model of a windmill near Grantham which, by all accounts, worked with the wind just like the real one.[61]

While Newton lived with Mr. Clark, he apparently fell in love with Mr. Clark's stepdaughter, Miss Storer, who revealed this story many years later. As far as we know, this is the only romantic relationship Newton ever had in his long life.[62]

Although Newton did nothing to distinguish himself while he was in Grantham School, he sufficiently impressed Mr. Stokes, the headmaster at Grantham School. Stokes interceded on his behalf to encourage Mrs. Newton to send him to the university rather than try to make a farmer of him, a career for which he had little desire and even less ability. His mother finally agreed, and in 1661, at 18, he went to Cambridge and enrolled at the famous and influential Trinity College. Because money was a problem, he became a Sizar, a student with money problems who pays his way through college by doing odd jobs and waits on other paying students and tutors.[63] At this time Cambridge was already 400 years old, initially established as a result of a migration from Oxford because of a crisis between the university an the town. It had grown to nearly five times the size of Oxford by the 1620s and long ago passed Oxford intellectually, as well.[64]

---

61 Richard S. Westfall, <u>Never At Rest</u>, p.60

62 Richard Westfall, <u>The Life of Isaac Newton</u>, p.13

63 Richard S. Westfall, <u>Never At Rest</u>, p.64.

64 Ibid, p.67.

Newton entered Trinity College of Cambridge University on June 5, 1661.[65] Assisted by his tutor, Benjamin Pulleyn, Newton embarked upon the standard curriculum of the time. By the end of his first three years, Newton had not distinguished himself in any way. By 1664 he faced a crisis because election to a scholarship was his only hope of continued residency at Trinity. Trinity only held elections to scholarships every four years, and the election in 1664 was the only one to be held during Newton's career as an undergraduate student. Newton did a very risky thing at that time; he abandoned the established and recognized course of instruction leading to a degree and took up an unorthodox course of his own design which had no standing whatever in the college. Fortunately, Pulleyn recognized Newton's genius and tried to help him by enlisting Isaac Barrow, who occupied the newly established Lucasian Chair of Mathematics, the position Newton would soon hold, and the one man in Trinity capable of judging Newton's competence. Even then it was difficult for Barrow because when he examined Newton on Euclid, Newton knew little or nothing because he had abandoned his study of Euclid after only a cursory review. Dr. Barrow did not examine him on Descartes' Geometry because he could not imagine that anyone could understand it without having first mastered Euclid. Although Newton was too modest to mention it, he had, in fact, mastered Descartes. There was something, however, that impressed Dr. Barrow sufficiently for him to recommend Newton and on April 24, 1664, Newton received his scholarship. There is only speculation as to a reasonable explanation for the decision. If Dr. Barrow was responsible for the recommendation and the resulting rescue from rural oblivion, as Richard Westfall calls it, it was

65 Richard Westfall, The Life of Isaac Newton, p.20

because he recognized a genius when he saw him, despite the geometry parts of the examination he had administered. Another possibility, and perhaps the more likely of the two, is Humphrey Babington, who had been a rector near Woolsthorpe and the brother of Mrs. Clark, the matriarch of the family with whom Newton had lived in Grantham. Babington was bursar of the college at the time, which may have proved once again that it is not who you know, but who knows you! In any event, Isaac Newton had a powerful advocate who has not been positively identified. If it was not either Barrow or Babington, it could be one more example of the many times when Isaac Newton most certainly had someone watching over his shoulder and protecting him. Under these circumstances, I will leave it to the reader to ponder considering the basic premise of this book.[66]

With the scholarship, Newton terminated his duties as a sizar. The scholarship provided him with a small income and, more importantly, the assurance of four more years of study, the probability of his Master of Arts degree, and the possibility of indefinite extension should he obtain a fellowship.

Isaac Newton's graduation with a Bachelor of Arts degree in 1665 ushered in a whole new chapter in the life of the world's greatest genius and an entirely new understanding of the workings of the universe for the rest of the world.

Newton's surviving notes indicate that he immediately plunged headlong into the study of mathematics. John Wickins, his roommate, remembers that when Newton began to work on a problem he would forget meals and ignore sleep. Wickens would find him the next morning completely self-satisfied at having discovered some proposition and totally

---

66 Richard S. Westfall, <u>Never At Rest</u>, p102

unconcerned with the loss of sleep. One of his contemporaries, John Whiston, remarked that "Newton could sometimes see the solution to a complex mathematics problem almost by intuition." Within a year Newton had digested all available text books and had moved on to an independent course of study. He began the exploration of solutions to curvatures which would soon become the calculus for which he is so famous, and the reason his picture appears in every calculus book everywhere.

In the summer after Newton's graduation, the plague descended on many parts of England, including Cambridge. On August 7, 1665, Trinity closed its doors and all the students dispersed into the country. The university would not reopen again until the spring of 1667.[67]

In Newton's own notes he records that in the beginning of 1665, several months before his graduation, he worked out the binomial theorem and by November, working by himself without any texts or guidance of any kind, had solved the method of fluxions, his initial title for differential calculus. By January, 1666, he had developed his theory of colors and by May had solved the inverse method of fluxions, integral calculus, and moved on into his theories of gravity and motion. During the two years of the plague he worked in Woolsthorpe, independently, with no books or other references, and before he was 24, he had invented calculus, both differential and integral, the Laws of Gravity, the Laws of Motion, and the Spectrum of Light—all original and independent works which changed the world forever. We can only imagine the number of meals and nights of sleep he lost during those two years.

It is significant that his first invention was calculus, the universal language that replaced the language lost at the Tower of Babel. Everything

---

67 Richard S. Westfall, <u>Never At Rest</u>, p.141

else Newton did is expressed in mathematical terms that everyone in the world can understand and provided a common language for people the world over to enable them to exchange ideas and opinions on the philosophical matters which explained the operation of our universe.

In early 1667, Newton returned to Trinity and was elected as a minor fellow and, a short time later, a major fellow. This identified him as an above average man in the college and provided him a good position. In the following year, Isaac Barrow resigned his position as the Lucasian Professor of Mathematics and arranged for Newton, at age 26, to succeed him.[68] That same year, Barrow published a book on optics and acknowledged Newton's help in the preface as a "man of quite exceptional ability and singular skill."

At this time, Newton worked on his own study of optics and actually devised and built machines for grinding lenses. He pioneered the use of pitch for optical polishing and succeeded in producing first-rate products for his research. In a dark room, he bored a small round hole to permit sunlight to penetrate and by arranging a prism in the beam of light, he discovered that the white light spread out into a colored strip of light in which the colors were red, orange, yellow, green, blue, indigo, and violet. He obviously had a very keen sense of color because few people can discern indigo as a distinct color between blue and violet. He reversed the test by bringing all the different colors together and overlapping and he proved that produced a light entirely and perfectly white. He then performed what he called the supreme test; he let a beam of white light fall on a prism and be spread out into spectrum. Then he took a board with a small hole in it and cut off all light except one color, perhaps red, and let

68 Richard S. Westfall, <u>Never At Rest</u>, p.102

the isolated red beam pass through a second prism. It was bent by the same amount as in the first prism and remained exactly the same color red. If the blue beam was let through to fall on the second prism, it was bent more than the red beams and remained the very same blue. This is how he demonstrated that each color has a discreet bending characteristic when light passes from air to glass. The amount of bending depends, of course, on the angle at which the beam strikes the surface. For a given angle, however, it is always greater for blue than for red.[69]

These experiments convinced Newton that a telescope could not be made using lenses that could provide a good image free of color bleeding. James Gregory, a Scottish mathematician and astronomer had, in 1663, proposed to solve that problem by using a mirror to construct a reflecting telescope, but he did not make it. Newton, with his own hands, constructed a mirror made of metal, a special alloy he prepared himself and he called it speculum metal (speculum is the Latin for mirror). He assembled the mirror into a telescope only six inches long but which magnified 40 times. Practically every telescope manufactured today is of the reflecting type and the largest telescopes, such as the 100-inch at Mount Wilson and the 200-inch at Mount Palomar, are reflecting telescopes. This is another example of Isaac Newton's contributions to the world, which are still current and used routinely by many people from amateur astronomists to scientists.

Edmund Halley visited Isaac Newton at Cambridge in 1684 to discuss a comet he had observed and asked Newton what he thought the curve of the comet's path would be. Newton answered immediately that it would be an ellipse. Halley was extremely pleased and asked how he knew, Newton replied, "I have calculated it". Then Halley asked if Newton

---

69 Richard Westfall, <u>The Life of Isaac Newton</u>, p.74

would teach him how to perform the calculus so he could calculate the path and predict its return. Newton agreed and Halley predicted it would return in 1758, years after the death of both men. It appeared on Christmas Day of that year and is known yet as "Halley's Comet". Could this, like his birth, be nothing more than another coincidence? Perhaps, but this marked one more solution to a riddle that had mystified the world for centuries. Sighting of the comet had been recorded all over the world for several hundred years before the birth of Christ. Perhaps the return of the comet on Christmas Day was God's way of acknowledging the achievement of Isaac Newton and Edmund Halley.[70]

Solutions to difficult and complex problems materialized to Newton by intuition. While he was at home during the plague he pondered the question of gravity. His orderly mind and intense concentration enabled him to progress from one proposition to another while retaining, for instant recall, a previous thought to fit into the overall scheme of things. It occurred to him that the power of gravity which brought an apple from the tree to the ground was not limited to a certain distance from the Earth and that it might extend as far as the moon. He reasoned that the same force that pulls the apple may also pull the moon. He reasoned that the moon revolving around the Earth was in a state of equilibrium between centrifugal force and it's opposite, a term he coined "centripetal force". Then he reasoned that the centripetal force, or gravity, must decrease as the distance from the Earth increases and deduced that it must weaken inversely as the square of the distance from the center of the Earth. (This is the Inverse Square Law) This extraordinary intellect devised laws of gravity and motion in 1665 that are still applicable today in our space program. The Inverse Square

70 Richard Westfall, The Life of Isaac Newton, p.160

Law is a major consideration in the computations necessary for the achievement of orbit of a space vehicle.

My research took me on a very winding path through the facilities of the National Aeronautical and Space Administration offices until I reached Gregory Oliver at the Johnson Space Center in Houston. He gathered two of his specialists, Barbara Conte and Philip Burley, and together they provided a thorough orientation and explained to me, at length, that the shuttle launches today are in exact accordance with Isaac Newton's Laws of Gravity and Laws of Motion. Virtually all the Laws of Gravity and Motion are demonstrated in every flight in both the ascent as well as the descent stages of the flight. The spacecraft is launched vertically (an object at rest tends to stay at rest) to allow it to rise up out of the densest part of the atmosphere at a lower speed to prevent it from burning or breaking up. The vertical path continues until the centrifugal force of the launching power is equalized by the centripetal force which tries to pull it down. The spacecraft is then pitched over horizontally and accelerated to orbital speed. (Without sufficient horizontal velocity it would simply fall back to Earth.) With the correct velocity, the spacecraft will begin to fall where the curvature of the Earth falls away, and the spacecraft will follow the curvature of the Earth as long as the velocity remains constant. (An object in motion tends to stay in motion.) When the mission is completed an on board rocket is fired in the direction of flight. This slows the spacecraft and velocity falls below that required to maintain orbit which results in the spacecraft falling back to Earth. In the vertical phase of the launch, the Laws of Gravity operate; in the horizontal orbit phase it is the Law of Motion that says that an object in motion tens to stay in motion.

It is also very interesting to note that Isaac Newton, in the early 1680s, drew the buttonhook trajectory that a spacecraft in orbit would take in its return path to Earth. This is the same sketch NASA used in the early days of space flight to show the general public the path of a returning orbiter each time it returned from its journey into outer space.[71]

Another story confirming the fact that Newton enjoyed the protection of God in so many ways involves the requirement for his ordination at Trinity College. A law had been on the books since Elizabeth's reign that made heresy a capital offense and included refusal to believe in the Trinity as one example of heresy. Newton did not believe in the Trinity, but avoided the attention of the authorities by merely being silent. They did not directly ask him and he did not talk about it. However, ordination was a necessary requirement to retain the Lucasian Chair of Mathematics at Trinity College, which he accepted in 1669, and this put a new perspective on the problem. He did not want to give up his position as the Lucasian Chair, not only because it was his livelihood, but also because it was his access to the world of science, which he so desperately needed. He believed that primitive Christianity—before Constantine established the Roman church in 313—to be the pure religion, and that the Roman church and the Pope had corrupted the religion by their idolatry. The Book of Revelation convinced him when it told: "If any shall worship the beast and his image and receive his mark upon his forehead or in his hand the same shall drink of the wine of the wrath of God." Newton did not doubt the truth of those words and therefore could not accept ordination. He had successfully evaded the requirement for several years, but by 1675 he had given up hope. In February, he made a trip to London to meet with the Secretary of the Royal Society to discuss a possible dispensation and, by April 27, an official dispensation announced that the Lucasian Chair of Mathematics would be exempt from the ordination obligation.[72] It never

71 Richard S. Westfall, <u>Never At Rest</u>, p.384.
72 Richard S. Westfall, <u>Never At Rest</u>, p.333.

mentioned Newton's name, only his position. Once again he was saved by an invisible hand. The preferred religion of the monarchs of England changed many times during Newton's lifetime and the punishment for disbelievers was swift and sure, but Newton managed to survive every change and every challenge which presented itself without a single scar. Someone had to be watching over his shoulder.

At about the time Newton began college in 1660, a charter from King Charles II established The Royal Society. In 1662 it was confirmed with the full title of "The Royal Society of London for the Promotion of Natural Knowledge." It is still the premier scientific society in the world today. Its motto, Nullius in Verba, freely translates as "We don't take anybody's word for it," Until that time, it was commonly accepted that any question could be answered by referring to the writings of Aristotle or some other Greek philosopher. Other detractors cited Descartes whom every educated man recognized as the greatest thinker of the time. The men who founded The Royal Society, however, believed in experimentation. Newton devoted a large part of his life to the Society, and a two-inch diameter reflecting telescope he made is still a treasured possession of the Society.

In his early 30s, Newton, after being well established in the scientific world for some time, began to weary of science. Robert Hooke, who was then secretary of The Royal Society, asked him to write a paper on the movement of the planets. Hooke had written a paper of his own on he subject that suggested the inverse square law of force, and an exchange of letters between the two had aroused Newton's interest. By 1684, with the question of planetary motions prominent in the discussions of the Society, Edmund Halley met with Christopher Wren and Robert Hooke to discuss it. Wren challenged Hooke to demonstrate the inverse square law and he could not do it. Then Halley went to Cambridge and asked for Newton's opinion on the motion of planets and the application of the inverse square law. Newton promptly provided two proofs of the elliptic orbit and the inverse square law.[73] Halley, recognizing the overwhelming importance of the proof, induced Newton to write a treatise setting out the details of his discoveries. And so the principia, heralded by men of science the world over as the greatest scientific book ever written, was born.

The Manuscript Philosophiae Naturalis Principia Mathematica, forever known and referred to the world over as the Principia, was presented and dedicated to the Royal Society on April 28, 1686. Although everything is contained in one book, it consists of three books or volumes, took about 18 months to write, and is derived from notes of experimental proofs which Newton made during the plague years when he was 24-25 years old. The Society could not financially afford to publish the document and so great credit must go to Edmund Halley for his efforts in production and expense of printing the book, Without his support, both personally and financially, his book might not have been published for some time.

In July, 1686, the President of The Royal Society was Samuel Pepys. Although not endowed with any kind of scientific knowledge in his own right, Pepys was very influential, well liked, and interested in the activities of learned men. He readily applied his name to the form signifying the approval of The Royal Society which constituted the necessary permission and sanction for the printing of the book. It was published in the summer of 1687. Halley, who financed the publication, sent 20 copies to Newton for his personal use and 40 more to the Cambridge bookstore with instructions regarding the selling price: he hoped to recoup his investment in the printing venture.

---

73 Richard S. Westfall, <u>Never At Rest</u>, p.403

The Principia, a mathematician's book for well-informed mathematicians, is not an easy book to read. Newton used methods which included analytical geometry, trigonometry and calculus for the explanation of his proofs. Renowned authors, such as Subrahmanyan Chandrasekhkar, the Nobel Prize winner in Physics, have produced magnificent works to simplify the understanding of problems and their solutions. Chandrasekhar had a profound admiration for a scientist whose work he believes to be unsurpassed and unsurpassable and attempted to bring into sharp focus the purity, majesty, and brilliant understanding of which Newton was possessed.

The Principia begins with due acknowledgement of Galileo's achievements and proceeds to prove Keplar's laws of planetary motion. In the first book, all the motions are considered as taking place in empty space with no resistance offered to the movement. In the second book, Newton deals with motions in surroundings that offer resistance such as bodies moving on or through water. And the third book is an absolute triumph. After summarizing in the first two, he proceeds to demonstrate the structure of the system of the world from the same principals. He did this with such thoroughness that what was achieved in the next 200 years by some of the most able-minded scientists was largely extensions and improvements of his work.

The Principia revealed the solutions to the riddles of the world while acknowledging the contributions of Galileo, Kepler, and others by scientific pronouncements so profound that the scientists and mathematicians of the world marveled at the depth of understanding demonstrated by the proofs offered in this book. No one ever even imagined an all-encompassing book of such intellectual depth, using original mathematical formulas developed specifically for a particular problem and using newly coined words such as centripetal, to be so profound that it immediately changed our understanding of the world and, indeed, the entire universe. Yet it was so simple: first he stated the rules of motion as if there was no resistance, then he proceeded to explain the effects of resistance, and then he topped it all off with an explanation of the structure of the universe.

One of the questions he resolves regards the shapes that meet with the least resistance. Consider the many applications of this in the design of ships operating on or under the surface of the water or airplanes operating at various altitudes and the concomitant varying densities of the air. Another application is the mathematical treatment of wave motion in which fundamental results were spelled out for the first time. Everyone knows today the significance of waves in modern science, but it was Newton who first explained it to us. He did the same thing with tides. As noted earlier in this chapter, he even worked out the familiar buttonhook path that an orbiting spaceship has to take to get back to Earth. We used to see this diagram in newspapers early in the space exploration of NASA. Now it has become so routine that no one thinks about these details any longer. We know the spaceship achieved orbit in the first place through the application of Newton's Laws of Gravity and Motion; but it also returns to Earth via the same rules.

The jet engine, which has become so common in aircraft and missiles, and more recently extended to boats and ships, is yet another example of the modern applications of one of Newton's Laws of Motion. And one has to think for only a moment to realize that without calculus none of the curved surfaces of airplanes or boats hull designs could have been formalized. Calculus, of course, has so many other applications to modern science that it affects practically everything in one way or another. Quality assurance is an integral part of every manufacturing process and uses calculus to establish probability tables for acceptance or rejection of a part or process. Calculus is used extensively in problem solving of statistical analysis, astronomy, electronics, nuclear physics, and space exploration. Computer design is another widely used application of calculus. The list of things, systems, and procedures which are affected by calculus goes on and on.

Isaac Newton did so many truly extraordinary things that it is difficult to imagine how one human mind could create all this original information in such a short time. He did all these things in about two years. For example, we know the axis of the Earth is tilted about 66½ degrees from the plane of orbit and that it always keeps parallel to itself in this orbit; this accounts for summer and winter. Actually, the Earth's axis does not keep precisely parallel to itself, but very slowly changes its angle so as to trace out a cone in the heavens—a movement which is called precession of the equinoxes. Very slowly may be an understatement because it takes nearly 26,000 years to complete the cone. If the Earth were a perfect sphere, Newton's principals would require that the force of the sun's attraction would pass exactly through the center and then the axis would always stay exactly parallel to itself. But the Earth is not a perfect sphere; it bulges slightly at the equator, and the pull of the sun on that part of the bulge nearest it is slightly greater than that on the part of the bulge farther away. This results in the production of a slight twist. Newton worked all this out and proved that the gravitational pull on the bulge would produce exactly the same very slow movement of the direction of the axis that had been observed by the "giants" for centuries. A calculation of this kind is not quite the same thing as suggesting that an apple falls because the Earth pulls it.[74]

Many similar explanations of this type make it easy for the reader to understand why the Principia is considered by the great men of science who came after Newton to be the most prodigious feat of human intellect. Laplace, the celebrated French mathematician and astronomer, said it best when he said, "The Principia is pre-eminent above any other production

---

74 E.N.da C. Andrade, <u>Sir Isaac Newton</u>, p.83.

of human genius. I say Isaac Newton was the object of a great deal of divine guidance!"[75]

For the first time in his life, at 54 years of age, Newton took up an entirely new lifestyle when he was appointed Warden of the Mint of England in the spring of 1696. He had lived at his beloved Trinity College in Cambridge for most of his life and only left there on short trips to London, primarily to attend meetings of The Royal Society. By 1699, he was elevated to Master of the Mint and his work became of great importance because clipping pieces off old coins in circulation had become popular and very profitable for the enterprising small-time crooks of London where the weight of some coins had been reduced by half. Newton, being a fine organizer and a good businessman, quickly developed branch mints with improved machinery and instilled efficiency into the coinage operations and record keeping. He held this extremely well paid post until his death.

During his years at the Mint, he did little scientific work, but retained his full analytical powers. Soon after his appointment to the Mint, he received two mathematical problems from the celebrated Swiss mathematician, Jean Bernoulli, as a challenge to "The acutest mathematicians of the world." One of the problems was, at that time, particularly difficult and required calculus for its solution. Newton worked them out that night before going to bed and gave a copy of the solutions to Charles Montagu, the President of The Royal Society. When Newton's solutions were sent to Bernoulli, without Newton's name, he exclaimed that he knew the solver at once from the style of his work—just as the lion is known by his claw.[76]

---

75 Ibid. p.86
76 Richard S. Westfall, <u>Never At Rest</u>, p.583

Newton was elected President of the Royal Society in 1703 and continued to take great interest in the Society's affairs until his death in 1727. He became the first distinguished man of science to hold the office since Christopher Wren gave it up more than twenty years earlier. And in that same year, 1703, another event took place which had a significant effect on Newton. In 1666, Newton had written his work on the Spectrum of Light and had run afoul of Robert Hooke, another English scientist, whose criticisms of Newton's works had irritated him immensely and had erupted into public disputes in 1675. Newton had vowed not to publish any more on the subject while Hooke lived. He died in 1703. Newton published his great work on light the following year. It was called "Opticks" and included with it two mathematical treatises on curves that he had previously written. The first part of the book deals with the experiments on the prism, and the colored images produced by lenses and the reflecting telescope. He also clearly explained the colors of the rainbow due to bending of light which penetrates into minute raindrops and, after reflection inside the drop, comes out again with a second refraction; the refraction by the water is accompanied by a separation of colors just as refraction by the glass prism. As a record of experiment and scientific deduction from experiment, Opticks is supreme.[77]

In 1705, Queen Anne and her court, including the Prince Consort George of Denmark, who was interested in science and had been elected a Fellow of The Royal Society the year before, visited Cambridge and made this the occasion of conferring a knighthood on Newton at Trinity College. He was not knighted for his services at the Mint but for his contributions and service to science and mankind. It was the first time

---

77 E. N. daC. Andrade, <u>Sir Isaac Newton</u>, p.106.

that such an honor had been conferred for achievements in pure science and it came as a clear sign and expression of the respect in which Newton was held.[78]

There were several other very important publications by Isaac Newton. One was De Analysi, a mathematical treatise related to calculus and printed in 1711. There was also Universal Arithmetic, not really arithmetic in the present sense of the word, but a treatise on algebra containing several discoveries of first importance. One of the most striking is the rule about equations, stated without proof, for which a proof was finally supplied more than 160 years later by the great mathematician, Sylvester.[79]

Perhaps equally important to his research, Sir Isaac Newton also established Rules of Reasoning and 15 Rules for Interpretation of Daniel in the Old Testament and the Book of Revelation in the New Testament and believed the two were really repetitious of each other. His interest and studies of all branches of religion consumed most of his adult life. The establishment of rules for interpretation was necessary in order to provide consistent meanings for the words in the text of the Books of Revelation which many times expressed the same thing in different terms.[80]

Shortly before his death he said, "I do not know what I may appear to the world, but to myself I seem to have been only like a boy, playing on the seashore and diverting myself in now and then finding a smoother pebble or a prettier seashell than ordinary, while the great ocean of truth lay all undiscovered before me."[81]

Sir Isaac Newton presided at the meeting of The Royal Society for the last time on March 2, 1727. He felt quite well after the meeting and stayed in London that night. Upon returning to his Kensington home

78 Richard S. Westfall, Never At Rest, p.624.

79 E. N. daC. Andrade, Sir Isaac Newton, p.118

80 Frank E. Manuel, The Religion of Isaac Newton, p.115

81 Richard S. Westfall, Never At Rest, p.863

the following day, he experienced severe pain and was diagnosed with kidney stones with no hope of recovery. He died on March 20, 1727, at 84 years old.

Newton believed, with his whole heart, that God had endowed him with the wisdom to interpret the Books of Revelation and solve the riddles of the universe. All of his studies were animated by one overwhelming desire, to know God's will through His works in the world. Newton believed also that a scientist who had the ability to explain the workings of the world and did not explain and share it with mankind, was denying God one form of adoration.[82]

Frank Manuel, in his book, "The Religion of Isaac Newton", emphasized that the fervor of Newton's quest for a knowledge of God was related to a psychic quest for his own father. In a fragment buried away in his church history, Newton proclaimed his submission to the Father. His confession of personal faith has a simple authenticity. "We must believe that there is one God or supreme Monarch that we may fear and obey him and keep his laws and give him honour and glory. We must believe that he is the father of whom are all things, and that he loves his people as his children that they may mutually love him and obey him as their father. We must believe that he is Lord of all things with an irresistible and boundless power and dominion that we may not hope to escape if we rebell and set up other Gods or transgress the laws of his monarchy, and that we may expect great rewards if we do his will. We must believe that he is the God of the Jews who created heaven and earth and all things therein as is exprest in the ten commandments that we thank him for our being and for all the blessings of this life, and forbear to take his name in vain or

---

82 Frank E Manuel, <u>The religion of Isaac Newton</u>, p.103.

worship images or other Gods. We are not forbidden to give the names of Gods to Angels and Kings, but we are forbidden to have them as Gods in our worship. For tho there be that are called God whether in heaven or in earth yet to us there is but one God the father of whom are all things and we in him and one Lord Jesus Christ by whom are all things and we by him: that is, but one God and one Lord in our worship."[83]

Alexander Pope, the greatest English poet of the early 1700s, paraphrased the passage from Genesis as an epitaph for Newton and said:
Nature and nature's laws lay hid in the night
God said "Let Newton be" and all was light.[84]

---

83 Ibid p.104.
84 Ibid p.20

Epilogue

The Industrial Revolution

Would the Industrial Revolution have occurred without the contributions of Isaac Newton? Perhaps a better question is *when* would the Industrial Revolution have occurred without Isaac Newton? If it had not occurred at the time it did, what would be the state of world technology today?

There is no doubt that the Industrial Revolution would have occurred with or without Isaac Newton; however, when it would have occurred is another matter. The number of scientific entities that Isaac Newton affected since his birth is truly amazing, and the scientific stature of the world today would be vastly different if Isaac Newton had not lived—or had not been endowed—with the God-given intellectual ability he possessed to solve the mysteries of the universe.

For several hundred years before the Industrial Revolution, only a few industries operated factories and assembly lines which produced uniform bricks, castings, and other smelting products. Patent laws and banks or money lenders necessary for financing technology also operated many years before the turn of the 16th century. Printed books helped to spread knowledge of technological developments, and the development of perspective became the milestone toward the improvement of drawings of machine parts.

The civil war of the 1640s began to change people's lives everywhere. The upheaval caused many people to question accepted thinking about religion, society, and nature. Once a civilization of agriculture, science soon crept in to influence the world, its people, and how they lived and worked. People began to consider their lives, society, politics, and religion. Many of the more disillusioned left England to start a new life in America. Hoping to set aside political and religious differences, small groups met to discuss scientific ideas. The Royal Society, one of the world's oldest scientific organizations, sprang from these endeavors in 1660, and new ideas were searched for around the world.[85]

By the 17th century, several inventors were experimenting with steam and soon recognized its potential as a source of power. The invention of writing, and printing with movable type, brought the development of the newspaper and for the first time in history, a great deal of knowledge was available to the common man on the street provided, of course, that he could read. Daily newspapers began to appear in London and invited the public to demonstrations of new or improved scientific instruments; it became a new form of entertainment for the common man.

Science, once a gentleman's diversion, developed into serious business by 1700. The Royal Society, of which Isaac Newton was president, provided a convenient meeting place and an encouraging environment for individuals to discuss their ideas and theories, as well as demonstrate their inventions.

Coffee houses sprang up in the city, and people like Stephen Demainbray, a well known speaker in London, lectured and demonstrated, using working models of the latest inventions and ideas. The coffee houses provided the common man with a convenient place to meet, exchange ideas, and learn about the newest technology and scientific

85 W. E. Lunt, <u>History of England</u>, p.566

inventions. The coffee houses were used by lecturers to explain Newton's theories in non-mathematical terms. The first public science lecture course, including Newton's works, was given in London in 1705.[86]

After the Copernican theory became accepted, elaborate instruments were collected by wealthy aristocrats who found science a fashionable and intellectual pastime. Mechanical models of the solar system, called "Orreries" became popular; they demonstrated the Copernican theory, which had the sun, rather than the Earth, at the center of the universe.

The Bank of London was founded in 1694 to provide financing for the developing economic system of commerce and trade. Towns such as Manchester and Leeds, began to grow up around factories and areas of industrial growth. However, economic growth became limited to local areas, primarily because of the extremely high cost of international business. The tremendous advantage of international trade was severely limited by the fact that inability to determine longitude at sea seriously impaired navigation, a problem that had persisted for more than four centuries and which affected every nation in the world which was engaged in or aspired to world trade.

Latitude was relatively easy to determine by observation of the North Star, but longitude was an entirely different matter. Many people, including Isaac Newton, believed that charts providing the positions of the moon and stars would provide the necessary information. But that was not to be. With no reliable charts or accurate methods of navigation, sailors had to rely on dead reckoning, which, at best, was merely a calculated estimate of the position of the ship in relation to land. Multitudes of stories are told of prolonged voyages resulting in an outbreak of scurvy or

---

86 Notes form British Museum of Science

some other disease that caused the deaths of thousands of sailors. All too often voyages ended in disaster when ships slammed against rocks due to an unexpected landfall. It is incredible that as recently as the middle 1700s there was no accurate method to determine longitude at sea.

One of the worst stories involved Sir Clowdisley Shovell who, returning home with his fleet of warships after a battle with the French in the Mediterranean, experienced twelve days of fog. Concerned that they might founder on coastal rocks, he asked all his navigators to put their heads together for a solution. Their consensus of opinion positioned the fleet safely west of Ile d' Ouessant, an island off the Brittany peninsula. But as they continued northward they discovered, to their horror, that they had misjudged their position which was actually near the Scilly Isles. These tiny islands lie about twenty miles from the southwest tip of England and point to land's end like stepping stones. That foggy night, October 22, 1707, the Scillies became unmarked tombstones for almost all of Sir Clowdisely's troops.[87] Similar scenarios, resulting in the deaths of many sailors, happened on voyage after voyage all over the world as ships crashed against shorelines during darkness, or became lost at sea. They also experienced outbreaks of scurvy and ran out of food. The Government of England, recognizing the desperate need for an accurate method to determine longitude at sea, passed the Longitude Act of 1714 which promised a cash prize of 20,000 pounds for a solution to the longitude problem.[88] Newton, who still served as president of The Royal Society, was appointed to serve as a judge on the longitude board and he pointed out that there were several possible methods of determining longitude at sea: a watch that kept perfect time, eclipses of the satellites

---

87 Dava Sobel, <u>Longitude</u>, p.11
88 Ibid, p 8.

of Jupiter, or the positions of the moon.[89] Many methods of determining longitude were proposed by people all over Europe. Some of the methods were limited to short distances from shore and most of the ideas were merely prize seekers who did not really have a viable solution. In 1773, John Harrison, the inventor of the chronometer, finally solved the problem.[90]

The solution to determining longitude at sea is knowing the time of day in Greenwich—the prime meridian. From the prime meridian it is 360° around the globe, which is also 24 hours. Dividing 360 by 24, we see that each hour is equal to 15° of longitude. At sea, the navigator resets his ship's clock each day to local noon when the sun reaches its highest point in the sky, then compares local noon to the Greenwich time, or 0° longitude. Then it is a simple matter to translate each hour difference to 15° of longitude. While we think of such a problem today as a simple and straight-forward comparison of two times, the solution was absolutely unattainable on the deck of a ship with a pendulum clock. These clocks would slow down or speed up as varying temperatures and humidity expanded or contracted the metal parts of the clock or thinned or thickened the lubricating oil. Even if they could be kept running, they still became virtually worthless.

It was John Harrison who came up with the solution: the chronometer. Harrison, a self taught genius, devoted his life to the perfection of his chronometer. Along the way, he invented two anti-friction aids which are still widely used today: the caged bearing and the jewel bearing. In test after test, his clocks kept extremely accurate time and the determination of longitude, a vexing problem that had provoked the world for centuries,

---

89 Richard S. Westfall, <u>Never At Rest</u>, p.834
90 Dava Sobel, <u>Longitude</u>, p.149.

was finally solved. In 1773, John Harrison finally received his award for his invention.

THE ROYAL OBSERVATORY
GREENWICH, ENGLSND

THE PRIME MERIDIAN
0° LONGITUDE
GREENWICH, ENGLAND

Harrison produced five chronometers, known as H-1 through H-5. Four of those clocks, H-1 through H-4 are displayed in The Royal Observatory at Greenwich. Today, three of them, H-1, 2, and 3 are still operating and are wound daily by the curator who refers to them reverently as "the Harrisons". H-4 lies operable but silent to preserve the masterpiece for the future. H-5 occupies a prominent position at The Clockmakers' Museum in Guildhall, London.[91]

With the obstacle of determining longitude at sea resolved, the development of trade on the high seas flourished along with the economic growth of the trading nations. The solution to the navigation problem that had escaped the learned nations of the world forever was most certainly one of the greatest contributions to the Industrial Revolution. It provided the means to develop accurate maps and charts of the world which had not previously been available, or even possible, and it is the basis for the time zones of the world which were taken for granted for so many years. World trade became much more efficient and accessible to developing nations, to say nothing of the lives it saved by preventing the type of accidents described in Dava Sobel's outstanding, true historical account, "Longitude".

In the meantime, Thomas Savery had patented his design in 1698 for a steam engine to raise water. By heating steam under great pressure in the boiler, then letting it into the cylinder, it forces water upwards. Using a jet of cold water the steam then condensed and the next load of water was let into the receiver from below. Savery engines were dangerous and inefficient but represented one of the first attempts to harness the power of steam.

---

91 Ibid, p146.

Selected by Isaac Newton, Francis Hawksbee served as curator of experiments to The Royal Society after the death of Robert Hooke. He created models of the members' inventions. With Newton's encouragement, he invented an air pump to create pressure that also reversed the pump action to create a vacuum. As an experiment, a burning candle or a small animal was put under the glass dome, and the air was pumped out, and the spectators were provided with the first proof that fire could not be sustained without air, and that animals would suffocate without air. During these experiments they also discovered that sound cannot travel in a vacuum. These are all very basic facts of science commonly known and taken for granted today, but only a little more than two hundred years ago, these things were totally unknown.

George Adams invented a pump in 1761 which was an improvement over the one previously developed by John Smeatum in 1750; Adams' pump could pump air out of a container as well as compress air into a container, but it could create higher pressures as well. This led to the discovery that compressed air could also be used to propel a bullet from a gun.

Fellows of The Royal Society experimented with new instruments, such as the air pump and the microscope, and reported their results in lectures and publications establishing conventions that are still followed today in creating scientific knowledge. The universal, rational laws of nature could, they believed, be uncovered by investigation and reason.

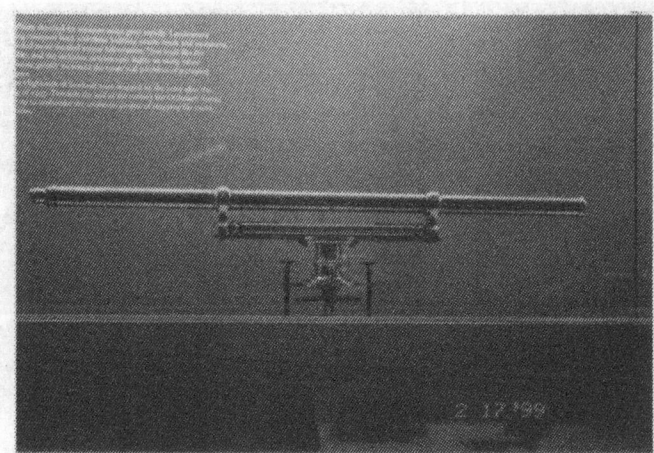

AN EARLY LEVEL INVENTED BY
JOSEPH JACKSON
MID 18TH CENTURY

AIR PUMP INVENTED BY FRANCES HAWKSBEE
1705

STEAM ENGINE PATENTED BY
THOMAS SAVERY

RAT'S TAIL CRANE

Measuring devices were invented which made it possible to greatly improve the accuracy of tools with the Vernier scale and micrometers, and also thermometers to measure heat, humidity, and air pressure. Alternative methods of transmitting motion or force, such as crank handles, drive belts, and foot treadles, were of great importance until the advent of electric motors—which did not happen until the late 19th century.

The so-called "rat's tail crane" invented in the 18th century, was commonly operated by men, horses or donkeys walking inside the wheel or sometimes by means of a rope outside the wheel, and became very popular for moving heavy cargo and masonry. The pile driver, an ingenious invention in 1737 by Richard Vauloue, who, ironically, had been a watchmaker. It was first used in the construction of the Westminister Bridge in 1738. Three horses would turn device's capstan that lifted a heavy weight which then, as it reached the desired height, activated a release mechanism causing the weight to fall on the pile.

In the middle 1700s, at about the same time that Joseph Jackson invented the level, Benjamin Cole invented the theodolite. These devices were critical to landowners, as well as to the military, for creating accurate maps, verifying property limits, locating and building roads, canals, and aqueducts. The level established and recorded various ground elevations encountered in a project, while the theodolite measured both vertical and horizontal angles.

James Watt designed more efficient steam engines in about 1770 and with his partner, Matthew Boulton, sold engines to many sectors of the industrial world. To encourage improvements in agriculture and industry, The Society for the Encouragement of Arts, Manufacturers, and Commerce was founded in 1754.

Encyclopedia Britannica, first published in 1787, showed old and new scientific devices at the end of the 18th century. This marked the first time that literature of this type was available for the general public. By the end of the 18th century the study of science had become specialized but, despite developments, there was very little formal science teaching and very few paid posts. For most people, scientific knowledge came from books, public lectures, and more and more through newly formed societies in individual towns such as Manchester, London, or Leeds. Many people were optimistic about progress and improvements that would result from science, while others, seeing the appalling conditions of the workers in the new factories, saw the darker side of scientific progress and wondered how it would affect the traditional powers of the church, state, and the monarchy. Today we see developments of science that include miraculous achievements in medicine, space exploration, and computer technology; there are biological biochemical, and biogenetic developments to provide better foods and nourishment. Improved construction equipment and methods provide better roads and bridges for our comfort; the pioneering exploration of the cosmos uses all of Newton's science to teach us more about the mysteries of our universe; and new metallurgical and composite materials discoveries make our planes, trains, and automobiles safer and faster for us to travel and enjoy the wonders of the far away lands of our planet. These are all examples of God's determination to provide the solutions to the riddles of the universe through Isaac Newton for our benefit.

All these wonderful things emanate from the discoveries and inventions of Isaac Newton, a 24 year old genius who lived less than 300 years ago and whose contributions to the science of the world continue to be the basis of new scientific achievements and discoveries as we enter the 21st century.

Sir Isaac Newton, the greatest scientist who ever lived, dedicated his life to understanding the workings of the universe and shared his findings with the world. Perhaps his most profound statement was "a scientist who has the ability to discover the secrets of the universe and fails to do so is denying one form of adoration to God.

It is difficult to tell what out lives might have been without the great advantage of God's gift to us: Isaac Newton.

Bibliography

American council of Learned Societies, Dictionary of Scientific Biography

Andrade, E.N. daC., *Sir Isaac Newton, His life and work*

Armstrong, Karen, *A History of God*

Beka Book Science Series, *Investigating God's World*

Beka Book Science Series, *Obseving God'Work*

Bible, The

Bragg, Melvyn, *On Ginats' Shoulders*

Branley, Franklin M., *Halley: Comet 1986*

Buck, Pearl S., *The Story Bible*

Bunch, Bryan, *Timetables of Technology*

Chandrasekhar, Subrahmanyan, *Confronting the Final Limit*

Chandrasekhar, Subrahmanyan, *Newton's Principia For the Common Reader*

Crystal, David, *The Cambridge Biographical Encyclopedia*

Edersheim, Alfred, *Bible History*

Encyclopedia of World Biography

Field, J. V., *Biography of Johannes Kepler*

Goldish, Matt, *Judaism in the Theology of Isaac Newton*

Grun, Bernard, *Timetables of History*

Guerlac, Henry, *Newton on the Continent*

Hadingham, Evan, *Early Man and the Cosmos*

Hart, Michael, *The 100*
Hunt, Garry, *Atlas of The Solar System*

Keynes, The Manuscripts, King's College, Cambridge

Lunt, W. E., *History of England*

Mailer, Norman, *The Gospel According to the Son*

Man, John, *Illustrated Encyclopedia of Astronomy*

Manuel, Frank E., *A Portrait of Isaac Newton*

Manuel, Frank E., *The Religion of Isaac Newton*

McMullin, Ernan, *Newton on Matter and Activity*

McNab, Andrew, *Newton, the Man*

Mott, Andrew, *The Principia Translated*

Pasachoff, Jay M., *Contemporary Astronomy*

Reston, James Jr., *Galileo, a Life*

Sagan, Carl, *The Pale Blue Dot*

Sobel, Dava, *Longitude*

Sobel, Dava, *Galileo's Daughter*

Stayer, Marcia Sweet, *Newton's Dream*

Voltaire, *Letters Concerning the English Nation*

Westfall, Richard, *The Life of Isaac Newton*

Westfall, Richard S. *Never At Rest*

Westminster Disctionary of Christian History

World Encyclopedia

CPSIA information can be obtained
at www.ICGtesting.com
Printed in the USA
LVHW031652261219
641757LV00015B/443/P

9 780595 129577